T0314023

Wastewater Treatment Technologies

Challenges in Water Management Series

Editor:

Justin Taberham
Publications and Environment Consultant, London, UK

Titles in the series:

Water Harvesting for Groundwater Management: Issues, Perspectives, Scope, and Challenges
Partha Sarathi Datta
2019
ISBN: 978-1-119-47190-5

Smart Water Technologies and Techniques: Data Capture and Analysis for Sustainable Water
Management
David A. Lloyd Owen
2018
ISBN: 978-1-119-07864-7

Handbook of Knowledge Management for Sustainable Water Systems
Meir Russ
2018
ISBN: 978-1-119-27163-5

Industrial Water Resource Management: Challenges and Opportunities for Corporate Water
Stewardship
Pradip K. Sengupta
2017
ISBN: 978-1-119-27250-2

Water Resources: A New Water Architecture
Alexander Lane, Michael Norton, and Sandra Ryan
2017
ISBN: 978-1-118-79390-9

Urban Water Security
Robert C. Brears
2016
ISBN: 978-1-119-13172-4

Wastewater Treatment Technologies

Design Considerations

Mritunjay Chaubey
Mumbai, India

WILEY Blackwell

This edition first published 2021
© 2021 John Wiley & Sons Ltd

All rights reserved. No part of this publication may be reproduced, stored in a retrieval system, or transmitted, in any form or by any means, electronic, mechanical, photocopying, recording or otherwise, except as permitted by law. Advice on how to obtain permission to reuse material from this title is available at http://www.wiley.com/go/permissions.

The right of Mritunjay Chaubey to be identified as the author of this work has been asserted in accordance with law.

Registered Offices
John Wiley & Sons, Inc., 111 River Street, Hoboken, NJ 07030, USA
John Wiley & Sons Ltd, The Atrium, Southern Gate, Chichester, West Sussex, PO19 8SQ, UK

Editorial Office
9600 Garsington Road, Oxford, OX4 2DQ, UK

For details of our global editorial offices, customer services, and more information about Wiley products visit us at www.wiley.com.

Wiley also publishes its books in a variety of electronic formats and by print-on-demand. Some content that appears in standard print versions of this book may not be available in other formats.

Limit of Liability/Disclaimer of Warranty
While the publisher and authors have used their best efforts in preparing this work, they make no representations or warranties with respect to the accuracy or completeness of the contents of this work and specifically disclaim all warranties, including without limitation any implied warranties of merchantability or fitness for a particular purpose. No warranty may be created or extended by sales representatives, written sales materials or promotional statements for this work. The fact that an organization, website, or product is referred to in this work as a citation and/or potential source of further information does not mean that the publisher and authors endorse the information or services the organization, website, or product may provide or recommendations it may make. This work is sold with the understanding that the publisher is not engaged in rendering professional services. The advice and strategies contained herein may not be suitable for your situation. You should consult with a specialist where appropriate. Further, readers should be aware that websites listed in this work may have changed or disappeared between when this work was written and when it is read. Neither the publisher nor authors shall be liable for any loss of profit or any other commercial damages, including but not limited to special, incidental, consequential, or other damages.

Library of Congress Cataloging-in-Publication Data

Names: Chaubey, Mritunjay, author.
Title: Wastewater treatment technologies : design considerations / Mritunjay Chaubey, Mumbai, India.
Description: Hoboken : Wiley-Blackwell, 2021. | Series: Challenges in water management series | Includes bibliographical references and index.
Identifiers: LCCN 2020032111 (print) | LCCN 2020032112 (ebook) | ISBN 9781119765226 (hardback) | ISBN 9781119765240 (adobe pdf) | ISBN 9781119765257 (epub)
Subjects: LCSH: Sewage–Purification. | Sewage disposal plants. | Water reuse.
Classification: LCC TD745 .C37 2021 (print) | LCC TD745 (ebook) | DDC 628.3–dc23
LC record available at https://lccn.loc.gov/2020032111
LC ebook record available at https://lccn.loc.gov/2020032112

Cover Design: Wiley
Cover Image: Maximilian Stock Ltd./Getty Images

Set in 9.5/12.5pt STIXTwoText by SPi Global, Pondicherry, India
Printed and bound by CPI Group (UK) Ltd, Croydon, CR0 4YY

10 9 8 7 6 5 4 3 2 1

Contents

Series Editor Foreword – Challenges in Water Management

The World Bank in 2014 noted:

> Water is one of the most basic human needs. With impacts on agriculture, education, energy, health, gender equity, and livelihood, water management underlies the most basic development challenges. Water is under unprecedented pressures as growing populations and economies demand more of it. Practically every development challenge of the 21st century – food security, managing rapid urbanization, energy security, environmental protection, adapting to climate change – requires urgent attention to water resources management.
>
> Yet already, groundwater is being depleted faster than it is being replenished and worsening water quality degrades the environment and adds to costs. The pressures on water resources are expected to worsen because of climate change. There is ample evidence that climate change will increase hydrologic variability, resulting in extreme weather events such as droughts floods, and major storms. It will continue to have a profound impact on economies, health, lives, and livelihoods. The poorest people will suffer most.

It is clear there are numerous challenges in water management in the 21st century. In the 20th century, most elements of water management had their own distinct set of organizations, skill sets, preferred approaches, and professionals. The overlying issue of industrial pollution of water resources was managed from a "point source" perspective.

However, it has become accepted that water management has to be seen from a holistic viewpoint and managed in an integrated manner. Our current key challenges include:

- The impact of climate change on water management, its many facets and challenges – extreme weather, developing resilience, storm-water management, future development, and risks to infrastructure
- Implementing river basin/watershed/catchment management in a way that is effective and deliverable
- Water management and food and energy security
- The policy, legislation, and regulatory framework that is required to rise to these challenges

- Social aspects of water management – equitable use and allocation of water resources, the potential for "water wars," stakeholder engagement, valuing water, and the ecosystems that depend upon it.

This series highlights cutting-edge material in the global water management sector from a practitioner as well as an academic viewpoint. The issues covered in this series are of critical interest to advanced-level undergraduates and Masters students as well as industry, investors, and the media.

Justin Taberham, CEnv
Series Editor
www.justintaberham.com

Preface and Acknowledgments

Globally, the practice of wastewater treatment before discharge does not seem to be good at this moment in time. As per the United Nations World Water Development Report 2017, on average, globally, over 80% of all wastewater is discharged without treatment. The discharge of untreated or inadequately treated wastewater into the environment results in the pollution of surface water, soil, and groundwater. As per the World Health Organization (WHO), water-related diseases kill around 2.2 million people globally each year, mostly children in developing countries. We need to understand that wastewater is not merely a water management issue – it affects the environment and all living beings, and can have direct impacts on economies.

The establishment of UN Sustainable Development Goal 6 (Clean Water and Sanitation) aims to ensure availability and sustainable management of water and sanitation for all, reflecting the increased attention on water and wastewater treatment issues in the global political agenda. The author of this book is convinced that water reuse is one of the most efficient, cost-effective, and ecofriendly ways to ensure water resilience. The author also believes that embedding sustainability into wastewater treatment is the best opportunity for industries to drive smarter innovation and efficient wastewater treatment. In order to develop sustainable wastewater treatment, we need to evaluate wastewater treatment systems in a broad sense. Economic aspects, treatment performance, carbon emissions, recycling, and social issues are important when evaluating the sustainability of a wastewater treatment system and selecting an appropriate system for a given condition. The modern concept of industrial wastewater treatment is moving away from conventional design. The trend is toward extreme modular design using smart and sustainable technology.

This book is intended as a reference book for all wastewater treatment professionals throughout the world. It may also be used as a textbook in the wastewater treatment training institutes and colleges and universities conducting graduate and postgraduate courses in the field of wastewater treatment and management. It will be equally useful for wastewater treatment plant operational personnel. This book covers a holistic view of the practical problems faced by the process industry and provides needs-based multiple solutions to tackle the wastewater treatment and management issues. This book elaborates on selection of right technology and design criteria for different types of wastewater. This will enable engineering students and professionals to expand their horizon in the wastewater treatment and management field.

The contents of this book are well organized. The book elaborates the problems and issues of wastewater treatment and then provides solutions and suggests how to implement those solutions in a sustainable way. Chapter 1 covers the global perspective of wastewater treatment, in which topics like global best practices, embedding sustainability into wastewater treatment, sustainable sources for industrial water, deep sea discharge, environmental rule of law, the polluter pays principle, and wastewater treatment technology trends are discussed. Chapter 2 is fully devoted to providing an understanding of the wastewater characteristics of several industries. In this chapter the author has tabulated the wastewater characteristics of more than 30 industries based on his personal and practical experience with these industries. Chapter 3 provides an overview of all commercially available wastewater treatment technologies. Chapter 4 details design considerations and design methodology, which are well illustrated based on the author's past 24 years of professional experience in wastewater treatment plant design. Chapter 5 covers the latest advances in sustainable wastewater treatment technologies. This chapter is purely based on the experience of the author of this book. The author has been involved in the piloting and customization of these technologies. Chapter 6 is devoted to zero liquid discharge technologies, offering an environmental and economic feasibility study. Chapter 7 covers wastewater treatment plant operational excellence. In this chapter, whether you are an experienced practitioner or an engineer who deals with the treatment of wastewater, you will find a myriad of practical advice and useful techniques that can immediately apply to solving problems in wastewater treatment. Throughout the book, attempts have been made to embed the theory in practical knowledge of wastewater technologies.

I extend my most sincere gratitude to my colleagues at Pentair, Shell, Unilever, and UPL for providing me with an environment to innovate and develop sustainable wastewater treatment technologies. I am also thankful to all my social media followers for encouraging me to write this book. The book could not have been completed without their motivation and support.

Last but not least, I would like to thank my family, especially Anubhav, Anushka, and Pummy, for providing moral support while writing this book.

Dr Mritunjay Chaubey
Ph.D. from Indian Institute of Technology Delhi

List of Abbreviations

ACF	activated carbon filter
AGF	activated glass media filter
AOP	advanced oxidation process
AOR	actual oxygen requirement
AOx	adsorbable organic halides
API	American plate interceptor
ASP	activated sludge process
ASU	activated sludge unit
ATFD	agitated thin film evaporator
BDD	boron-doped diamond
BOD	biochemical oxygen demand
BP	belt press
BTEX	benzene, toluene, ethylbenzene, and xylene
BW	brackish water
CAGR	compound annual growth rate
CAPEX	capital expenditures
CAS	conventional activated sludge
CEB	chemically enhanced backwash
CETP	common effluent treatment plant
CIP	cleaning in place
COC	cycle of concentration
COD	chemical oxygen demand
conc.	concentrated
CWAO	catalytic wet air oxidation
DAF	dissolved air flotation
DI	deionized
DMF	dual media filter
DNB	denitrification/nitrification biotreater
DO	dissolved oxygen
EBT	eriochrome black T
EDTA	ethylene diamine tetra acetic acid
EO	electro-oxidation
EPDM	ethylene propylene diene monomer

ETP	effluent treatment plant
F/M	food to microorganism (ratio)
FAS	ferrous ammonium sulfate
FFBR	fixed film biological reactor
FIC	flow indicator control
FMCG	fast moving consumer goods
FO	forward osmosis
FP	filter press
GAC	granular activated carbon
HAL	*hybrid anerobic lagoon*
HRSCC	high rate solid contact clarifier
HTDS	high total dissolved solids
HRT	hydraulic retention time
I&C	instrumentation and control
INDB	intermittent nitrification/denitrification biotreater
IWRM	integrated water resources management
LIC	level indicating control
LIT	level indicating transmitter
LSI	Langlier Saturation Index
LTDS	low total dissolved solids
MBBR	moving bed biofilm reactor
MBR	membrane bioreactor
MEE	multiple-effect evaporator
MGF	multigrade filter
MLSS	mixed liquor suspended solids
MLVSS	mixed liquor volatile suspended solids
MMO	mixed metal oxide
MS	mild steel coated with epoxy paint
NIST	National Institute of Standards and Technology
NOM	natural organic matter
NTU	nephelometric turbidity units
O&G	oil and grease
OPEX	operating expenses
ORP	oxidation reduction potential
PABR	*packed anerobic bed reactor*
PLC	programmable logic controller
PM	preventive maintenance
ppm	parts per million
PSF	pressure sand filter
PVDF	polyvinylidene fluoride
RBC	rotating biological contractor
RCC	reinforced cement concrete
RO	reverse osmosis
SBR	sequencing batch reactor
SDG	Sustainable Development Goal

SND	simultaneous nitrification and denitrification
SOP	standard operating procedure
SRT	sludge retention time
SSVI	settleable sludge volume index
SVI	sludge volume index
TDS	total dissolved solids
TKN	total Kjeldahl nitrogen
TOC	total organic carbon
TPI	tilted plate interceptor
TSS	total suspended solids
UN	United Nations
UNEP	United Nations Environment Program
UASB	up-flow anerobic sludge blanket
UF	ultrafiltration
US-EPA	United States Environmental Protection Agency
UV	ultraviolet
VFD	variable frequency driver
VOC	volatile organic compounds
WAS	waste activated sludge
WHO	World Health Organization
WW	wastewater
WWTP	wastewater treatment plant
ZLD	zero liquid discharge

1

Global Perspective of Wastewater Treatment

1.1 Global Wastewater Treatment Scenario

Natural water in contact with foreign matter during either industrial manufacturing processes or domestic use becomes polluted. Such polluted water is termed wastewater. The removal of excessively accumulated foreign matter from wastewater is known as treatment. As is well known, the rate at which we deplete and degrade our fresh aquatic resources poses a great threat to our future life-support system. The rise in human population exploits more natural resources and this is met through the growth of industries, urbanization, deforestation, and intensive agricultural practices. Industries and urban sprawl discharge waste into rivers, the deforestation process itself aggravates sedimentation transport into streams, and the use of chemicals contaminates groundwater through percolation and rivers and lakes through surface run-off.

All these sporadic degrading activities have led to gradual deterioration in the quality of surface and subsurface water. The loss of water quality is causing health hazards, death of human-beings, death of aquatic life, crop failures, and loss of esthetics. Keeping in mind these alarming global problems and the importance of environmental and nature protection, the 1972 Stockholm Conference on the Human Environment was the first of its kind and jolted the world into an awareness of environmental issues. A tangible result of that conference was the setting up of the United Nations Environment Program (UNEP) to serve as the conscience of the UN in matters concerning the environment. Twenty years later came the next landmark – the Earth Summit in Rio de

Wastewater Treatment Technologies: Design Considerations, First Edition. Mritunjay Chaubey.
© 2021 John Wiley & Sons Ltd. Published 2021 by John Wiley & Sons Ltd.

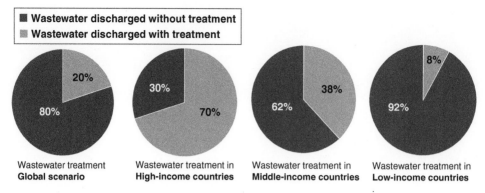

Figure 1.1 Global wastewater treatment scenarios.

Janeiro in 1992 – which exceeded everybody's expectations in terms of the number of attendees and scope of topics discussed. The Summit's message was broadcast to the world: "that nothing less than a transformation of our attitudes and behavior would bring about the necessary changes."

Globally, the practice of wastewater treatment before discharge does not seem to be good. On average, high-income countries treat about 70% of the municipal and industrial waste-water they generate [1]. That percentage drops to 38% in upper middle-income countries and to 28% in lower middle-income countries. In low-income countries, only 8% of waste-water undergoes treatment of any kind. These estimates support the often-cited approximation that, globally, over 80% of all wastewater is discharged without treatment. In high-income countries, the motivation for advanced wastewater treatment is either to maintain environmental quality or to provide an alternative water source when coping with water scarcity. However, the release of untreated wastewater remains common practice, especially in developing countries, due to lack of infrastructure, technical and institutional capacity, and financing (see Figure 1.1).

The discharge of untreated or inadequately treated wastewater into the environment results in the pollution of surface water, soil, and groundwater. The effects of releasing untreated or inadequately treated wastewater can be classified with regard to three issues:

- Adverse human health effects.
- Negative environmental effects due to the degradation of water bodies and ecosystems.
- Potential effects on economic activities: as the availability of freshwater is critical to sus-tain economic activities, poor water quality constitutes an additional obstacle to economic development.

According to the World Health Organization (WHO), water-related diseases kill around 2.2 million people globally each year – mostly children in developing countries.

1.2   The UN Sustainable Development Agenda
for Wastewater

In September 2015, approximately 193 nation members of the United Nations General Assembly unanimously adopted "Agenda 2030" with a total of 17 Sustainable Development Goals (SDGs) to end poverty, protect the planet, and ensure prosperity for all (see Figure 1.2).

Figure 1.2   UN Sustainable Development Goals.

The establishment of SDG 6 (Clean Water and Sanitation) is aimed at ensuring availability and sustainable management of water and sanitation for all, reflecting the increased attention on water and wastewater treatment issues in the global political agenda. Agenda 2030 lists rising inequalities, natural resource depletion, environmental degradation, and climate change as among the greatest challenges of our time. It recognizes that social development and economic prosperity depend on the sustainable management of freshwater resources and ecosystems and it highlights the integrated nature of SDGs.

SDG 6 includes eight global targets that are universally applicable and aspirational. SDG 6 covers the entire water cycle, including: provision of drinking water (target 6.1) and sanitation and hygiene services (6.2); improved water quality, wastewater treatment, and safe reuse (6.3); water-use efficiency and scarcity (6.4); integrated water resources management (IWRM) including through transboundary cooperation (6.5); protecting and restoring water-related ecosystems (6.6); international cooperation and capacity-building (6.a); and participation in water and sanitation management (6.b).

SDG target 6.3 (to improve water quality, wastewater treatment, and safe reuse) focuses mainly on collecting, treating, and reusing wastewater from households and industry, reducing diffuse pollution and improving water quality. As per SDG 6 Synthesis Report 2018 on Water and Sanitation [2], ambient freshwater quality is at risk globally. Freshwater pollution is prevalent and increasing in many regions worldwide. Preliminary estimates of household wastewater flows from 79 mostly high- and high-middle-income countries show that 59% is safely treated. For these countries, it is further estimated that safe treatment levels of household wastewater flows with sewer connections and on-site facilities are 76 and 18%, respectively.

The degree of industrial pollution is not known, as discharges are ineffectively observed and only from time to time calculated and aggregated at national level. Although some local and modern wastewater is treated nearby, hardly any information is accessible and amassed for national and territorial evaluations. Numerous nations come up short on the ability to gather and analyse the information required for a full appraisal. Reliable water quality monitoring is fundamental to the direct needs for ventures. It is also important for assessing the status of aquatic ecosystems and the need for protection and restoration.

Increasing political will to tackle pollution at its source and to treat wastewater will protect public health and the environment, mitigate the costly impact of pollution, and increase the availability of water resources. Wastewater is an undervalued source of water, energy, nutrients, and other recoverable by-products. Recycling, reusing, and recovering what is normally seen as waste can alleviate water stress and provide many social, economic, and environmental benefits.

Managing wastewater by implementing global best practices of wastewater collection and treatment can support achievement of SGD target 6.3. Wastewater should be seen as a sustainable source of water, energy, nutrients, and other recoverable by-products, rather than as a burden. Choosing the most appropriate type of wastewater treatment system that can provide the most co-benefits is site specific, and countries need to build capacity to assess this. Reuse of water needs to take into account the whole river basin, as wastewater from one part of a basin may well be the source of supply for others downstream.

Managing wastewater and water quality also needs to include better knowledge of pollution sources. SDG reporting could support countries in aggregating wastewater subnational data and publicly reporting at the national level. This would include monitoring performance to ensure treatment plants are managed and maintained to deliver effluent suitable for safe disposal or use according to national standards, which may vary from country to country. Countries that do not have national standards and monitoring systems need to assess performance of on-site and off-site domestic wastewater treatment systems. Formalizing the informal sector through various policy instruments is needed to prevent excessive contamination. Incentives for the informal sector to be registered with the government could be accompanied by combined analysis of all wastewater sources and their relative contribution to health and environmental risks. This would enable countries to prioritize investments in pollution control that contribute most to achieving SDG target 6.3.

1.3 Global Market Size

The global market for water and wastewater technologies reached USD 64.4 billion in 2018 and should reach $83.0 billion by 2023, at a compound annual growth rate (CAGR) of 5.2% for the period 2018–2023 [3] (see Figure 1.3).

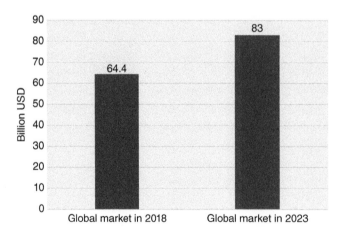

Figure 1.3   The global market size of water and wastewater technologies.

1.4 Global Best Practices

Treatment of wastewater has received steadily increasing attention across the world. During the manufacturing of industrial products, wastewater is generated at various stages which is very complex in nature and highly variable in quantity and quality.

Unless we adopt a structured approach toward collection, segregation, and treatment, it will be difficult to achieve the desired results. Best practices in wastewater treatment include:

- Reducing water consumption at source.
- Maximizing recycling and reusing during production.
- Promoting effluent identification, characterization, and segregation at source.
- Deploying sustainable technology to treat wastewater.
- Minimizing treated wastewater disposal.
- Eliminating incineration of wastewater.

See Figure 1.4.

A structured approach toward wastewater treatment help us to manage complex industrial effluent.

Figure 1.4 Best practices in wastewater treatment.

1.4.1 Effective Wastewater Treatment

Wastewater treatment is a complex process, and a properly operated wastewater treatment plant has many requirements. Below are six top considerations for effective wastewater treatment.

1.4.1.1 Discharge Standards

The first consideration is to understand the local discharge standards from the environmental authority. The authority may require that you submit a permit application or notice of intent that typically describes the sources, characteristics, and volumetric flow of your industrial wastewater discharge.

1.4.1.2 Wastewater Inlet Characteristics

We should understand the processes that produce waste streams and the wastewater characteristics of each stream. Review procedures for how products and reagents are combined to produce wastewater streams. Once we have sound knowledge of the characteristics and variability of the wastewater, we can design a treatment system and develop protocols to ensure continuous and compliant operation.

1.4.1.3 Wastewater Mass Balance

We should be familiar with the mass balance of how much water flows into a manufacturing plant and how many pollutants are in the wastewater. By conducting mass balances on all the constituents, a thorough understanding of the process can be obtained, leading to the optimal performance of the system. Flow rate is the most critical factor when calculating the capacity of a wastewater treatment system.

1.4.1.4 Wastewater Segregation

A structured approach toward wastewater treatment help us to manage complex industrial effluent. The best way to manage complex and variable industrial wastewater is through wastewater stream identification, characterization, and segregation (see Figure 1.5). The all-incoming effluent stream should be identified and segregated into green, yellow, and red streams. The green stream may consist of all the wastewater stream having total dissolved solids (TDS)<5000 ppm and chemical oxygen demand (COD)<10000 ppm. The yellow stream may consist of all the wastewater stream having TDS<100000 ppm and COD<20000 ppm. The red stream may consist of all the wastewater stream having TDS>100000 ppm and COD>20000 ppm. After stream identification and segregation, the green stream may be treated with biological treatment technologies such as an activated sludge process or a moving-bed biological reactor; the yellow stream may be treated with forward osmosis (FO), Scaleban, or OH radical technology; and the red stream may be treated with multi-effect evaporation technology or any other appropriate evaporation technology.

Green stream	Yellow stream	Red stream
TDS<5000 ppm COD<10 000 ppm	5000<TDS<100 000 ppm COD<20 000 ppm	TDS>100 000 ppm COD>20 000 ppm
Biological treatment is best treatment solution for green stream effluent	Advanced oxidation and forward osmosis are best treatment solutions for yellow stream effluent	Advanced close evaporation is best treatment solution for red stream effluent

Figure 1.5 Wastewater stream segregation.

1.4.1.5 Sustainable Technology

We should use sustainable wastewater treatment technologies that consume less power and chemicals, generate less hazardous solid waste, and use minimum manpower. Use of sustainable wastewater treatment technology is the best opportunity for industries to drive smarter innovation and efficient wastewater treatment. Sustainable technology ensures a pollution-free society, compliance with environmental norms, and creation of wealth from waste.

1.4.1.6 Standard Operating Procedures

Standard operating procedures (SOPs) of wastewater treatment plants must be documented and available to operators for reference. It is important for operators to know their daily, weekly, and monthly responsibilities. Operators are a key resource in the wastewater treatment plant. Operators are responsible for managing pumps, probes, and filtration equipment, general housekeeping, testing alarms, and any other tasks to keep a safe and orderly facility. If new technologies are added to the system, operators must be trained to operate these.

1.5 Embedding Sustainability into Wastewater Treatment

Embedding sustainability into wastewater treatment provides the best opportunity for industries to drive smarter innovation and efficient wastewater treatment. To analyze the full scope of embedding sustainability into wastewater treatment we can use a lifecycle cost analysis tool. To reach a final decision, the different indicators should be normalized and weighted to integrate them into a single final objective, which makes the search for a sustainable solution a multi-objective optimization problem. Important objectives in selecting sustainable wastewater treatment technologies are as follows:

- Minimize use of resources such as water, energy, chemicals, and space.
- Minimize treatment costs.
- Minimize production of harmful waste products.
- Minimize use of manpower.
- Maximize the treatment efficiency.
- Maximize social-cultural embedding through acceptance, participation, and stimulation of sustainable behavior.

Due to the complexity and the dynamic understanding of today's problems, there is a risk of introducing new problems when implementing technical solutions. To ensure that solutions have a positive overall impact on society, one needs to be clear about lifecycle cost assessment.

In order to develop sustainable wastewater treatment, we need to evaluate wastewater treatment systems in a broad sense. Economic aspects, treatment performance, carbon emissions, recycling, and social issues are important when evaluating the sustainability of a wastewater treatment system and selecting an appropriate system for a given condition. Selection of a wastewater treatment scheme requires a multidisciplinary approach in which engineers and technocrats discuss with economists, biologists, health officials, and the public.

1.5.1 Optimizing the Operating Cost of Wastewater Treatment Plants

Under sustainability, optimizing the operating cost and reducing the environmental footprint of a wastewater treatment plant are very important. Here, one case study of the wastewater treatment cost of a chemical industry is presented to understand the actual cost associated with wastewater treatment inside a large chemical manufacturing plant (see Table 1.1). Based on actual data received from the wastewater treatment plants of various chemical industries, the author of this book summarizes the average cost of wastewater treatment within chemical and agrochemical manufacturing plants. This operating cost is for a biological wastewater treatment plant to treat effluent having an inlet COD of 5000–8000 ppm and treated outlet COD <250 ppm (see Figure 1.6).

Table 1.1 Average cost of wastewater treatment plants in chemical industries.

Cost monitoring parameters	Average cost (USD/m³)
Power cost	1
Chemical cost	1.5
Sludge disposal cost	1
Treated effluent disposal cost	0.5
Manpower cost	1
Total cost	**5**

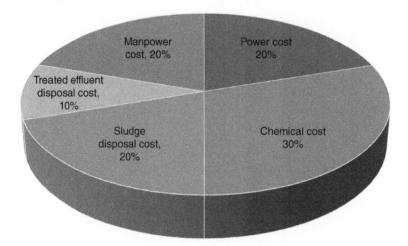

Figure 1.6 Operating cost composition of chemical industry wastewater treatment plants.

With the use of sustainable wastewater treatment technologies, we may further reduce power consumption, chemical consumption, sludge generation, and involvement of man-power in wastewater treatment plants.

1.5.2 New Sustainable Wastewater Treatment Technologies

Use of sustainable wastewater treatment technologies is a key factor in embedding sustainability into wastewater treatment. Sustainable wastewater treatment systems depend on a number of factors including minimal use of resources such as water, energy, chemicals, and space; minimum treatment costs; minimum production of harmful waste products; minimum use of manpower; maximum treatment efficiency; and maximum social-cultural embedding through acceptance, participation, and stimulation of sustainable behavior. Some of these new sustainable wastewater treatment technologies are now discussed.

1.5.2.1 Forward Osmosis

FO is a membrane-based wastewater treatment technology utilizing drawdown solution to treat high TDS (<100 000 ppm) and moderate COD (<20 000 ppm). FO is a natural process and an integral part of the survival of flora and fauna on this planet. In general, the FO process is governed by differences in osmotic pressure, and the direction of water diffusion takes place from a lower concentration (the feed side) to a higher concentration (the draw side). The driving force for this separation is an osmotic pressure gradient which is generated by a draw solution of high concentration to induce a net flow of water through the membrane into the draw solution, thus effectively separating the feed water from its solute. As osmosis is a natural phenomenon, it significantly requires less energy compared to the conventional reverse osmosis (RO) process. FO technology can be used for highly saline waters which are impossible to treat through conventional wastewater treatment processes (see Figure 5.6).

1.5.2.2 Scaleban

Scaleban is a unique and patented technology that helps industries achieve water conservation and zero liquid discharge (ZLD) by integrating process effluent and RO reject water having high TDS with existing cooling towers in place of freshwater. Scaleban uses a cooling tower as a natural evaporator without affecting the plant's performance in relation to hard water scaling, corrosion, and bio-fouling in the cooling tower circuit. With application of the Scaleban system, cooling towers can be operated at higher TDS; hence effluent treatment plant (ETP)-treated water/effluent can be used as the cooling tower makeup water, thus reducing raw water consumption without requiring any extra energy input for its operation (see Figure 1.7).

1.5.2.2.1 Advantages

- Reduced abstracted water demand in the cooling tower by utilizing treated wastewater.
- Much less capital and operational expenditure compared to conventional technologies to achieve ZLD.

Figure 1.7 Scaleban equipment installed in a cooling tower.

Figure 1.8 Volute press equipment.

- Can handle higher COD and TDS water efficiently.
- Quick installation and commissioning without occupying any extra footprint.
- No major infrastructural changes required for installation.

1.5.2.3 Volute Press
A volute press is a multidisc sludge dewatering press that removes water and moisture from sludge on a continuous basis. It consists of two types of rings: a fixed ring and a moving ring. A screw tightens the rings and pressurizes the sludge. Gaps between the rings and the screw are designed to gradually get narrower toward the direction of the sludge cake outlet, and the inner pressure of the discs increases due to the volume compression effect, thickening and dewatering the sludge (see Figure 1.8).

1.5.2.3.1 Advantages

- Continuous and clean operation without regular manual intervention.
- Produces high-quality filtrate with much less total suspended solids (TSS) (i.e. high solid recovery).
- Extremely low power consumption – reduces power consumption up to 95%.
- Low noise and odor generation.
- Low wash water consumption.

1.5.2.4 Moving Bed Biofilm Reactor
The moving bed biofilm reactor (MBBR) system is an advanced activated sludge process whereby biological sludge is immobilized on plastic carriers having a very large internal surface area. The aeration system keeps the carriers with activated sludge in motion, thus providing a larger and wider contact between microorganisms and wastewater for efficient wastewater treatment (see Figures 3.13 and 1.9).

Figure 1.9 An MBBR plant.

1.5.2.4.1 Advantages

- Compact system with smaller area footprint compared to conventional activated sludge process.
- Higher food to microorganisms ration (F/M) loading with reduced retention time.
- Less biological sludge generation and no biomass recycling required.
- Faster installation and commissioning.
- Higher treatment efficiency.

1.5.2.5 Dissolved Air Floatation

Dissolved air floatation (DAF) technology is a modern version of conventional primary effluent treatment, where suspended solids are removed by dissolving atmospheric air in wastewater under pressure and then releasing the air in a flotation tank basin. The released air forms tiny bubbles, causing the suspended matter to float on the surface, and in turn can be removed from wastewater using a skimming device (see Figure 3.6).

1.5.2.5.1 Advantages

- Very compact system which reduces the area footprint significantly.
- Quick installation and commissioning.
- Higher suspended solids removal efficiency with ability to handle bulking floating solids.
- Lower capital expenditures (CAPEX) and operating expenses (OPEX).

1.6   Sustainable Sources for Industrial Water

Industrial water scarcity is one of the major impacts on businesses worldwide, leading to higher operating costs and difficulty in staying competitive. For industries, day by day controlling costs is difficult and this worsens when the price of water increases exponentially to the point where profit margins shrink precariously. This causes industries to regard water access as a competitive advantage and to adopt sustainable sources for industrial water. In the following sections, the various sources of industrial water along with approximate costs are summarized.

1.6.1 ZLD Water

ZLD water is generated inside industrial manufacturing plants by adopting ZLD treatment methods. Industry uses a number of technologies and various stages of wastewater treatment to get ZLD water. The approximate cost of ZLD water is in the range of USD 10–17/ m^3 water produced. ZLD water not only is costly but also generates huge amounts of carbon and hazardous solid waste. Due to the higher cost and higher environmental footprint, ZLD water is not a sustainable source for industrial water.

1.6.2 Desalinated Water

Desalinated water is generated by desalination of seawater by adopting RO treatment methods. Industry uses various kinds of membrane to produce desalinated water. The approximate cost of desalinated water is in the range of USD 0.7–1/m^3 water produced. In coastal areas, desalinated water seems to be a sustainable source for industrial water.

1.6.3 Sewage Water

Sewage water is generated from municipal sewage treatment plants by treating domestic sewage. This treated domestic sewage will be further treated by industries as per requirement . Sewage generation is increasing rapidly on a global basis, and in the absence of adequate infrastructure for collection and treatment, the already depleting freshwater reservoirs are being polluted. In the present global water scarcity, *sewage wastewater is the new black gold on the planet Earth*. Various decentralized sewage treatment facilities are being set up for recycling and reuse of wastewater. Advanced treatment technologies are being adopted for sewage treatment. Globally, there is increasing focus on adding treatment capacity, improving collection efficiency, and automating operations for wastewater. New public–private partnership models and long-term operations and maintenance contracts are being introduced to benefit wastewater treatment plants. These measures will improve wastewater management and generate social, environmental, and economic benefits, and are essential to achieving Agenda 2030 SDGs.

The approximate cost of sewage water is in the range of USD 0.4–0.5/m^3 water produced. In areas where a sufficient amount of treated municipal sewage is available for industrial use, sewage water seems to be a sustainable and economical source for industrial water. By using sewage water in industrial manufacturing, we can also prevent water pollution.

1.6.4 Rainwater

During rainy seasons we get huge amounts of rainwater. We need to collect, store, filter, and reuse rainwater for manufacturing processes. Industrial rainwater is becoming increasingly important for commercial entities to reduce their environmental impact across their operations. Industrial rainwater harvesting is an extremely cost-effective method of achieving this goal, with the added benefit of reducing water consumption and bills. Industrial rainwater harvesting systems are easy to install and maintain, whilst providing

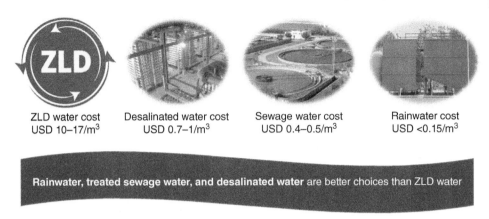

| ZLD water cost
USD 10–17/m³ | Desalinated water cost
USD 0.7–1/m³ | Sewage water cost
USD 0.4–0.5/m³ | Rainwater cost
USD <0.15/m³ |

Rainwater, treated sewage water, and desalinated water are better choices than ZLD water

Figure 1.10 Alternative sources for industrial water.

cost-effective savings on water consumption; resulting in reduced water bills. The approximate cost of rainwater harvesting is less than USD $0.15/m^3$ water produced. In areas where a sufficient amount of rainfall is available, rainwater is thus the better and sustainable choice of all available water options (see Figure 1.10).

1.7 Deep Sea Discharge as an Alternative to Minimize Human and Environmental Health Risks

Deep sea discharge of treated wastewater can be an effective, reliable, and economical solution to wastewater disposal that has minimal environmental impacts and avoids water pollution problems in coastal regions. The marine environment has a high capacity for dispersion and decay of organic matter. This capacity lies in the available energy in the marine environment due to the action of ocean currents on wastewater dispersion, the availability of dissolved oxygen, and due to it being a hostile environment to the survival of microorganisms.

The National Research Council of the US National Academy [4] specifically recommended against a "one size fits all" approach to arbitrary specification of treatment levels, stating:

> Coastal wastewater and stormwater management strategies should be tailored to the characteristics, values, and uses of the particular receiving environment based on a determination of what combination of control measures can effectively achieve water and sediment quality objectives.
>
> Sydney's deepwater ocean outfalls have delivered high-quality outcomes for the environment and the community [5]. Beaches and harbors are cleaner and the marine environment is healthy.

Since the deepwater ocean outfalls opened in 2004,

- Swimming conditions have significantly improved.
- Beach grease has been eliminated.
- There has been no detectable negative effect on marine ecology or sediments.
- Effluent discharged has consistently been shown to be non-toxic at its diluted state.

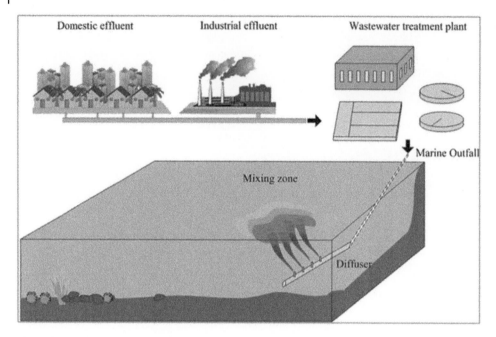

Figure 1.11 A deep sea discharge wastewater disposal system.

A typical deep sea discharge system for treated wastewater disposal is shown in Figure 1.11. It usually consists of a wastewater collection pipeline, combined wastewater treatment plant, and discharge structure – the deep sea discharge outfall.

Deep sea discharge outfalls release treated wastewater 2–10 km off the coast, where it mixes with seawater with the help of diffusers. The primary treated effluent is conveyed through tunnels under the ocean floor and is released through a series of diffusers. These diffusers release the effluent in fine jet streams, so it mixes immediately with seawater and disperses into the strong sea current. Because it is less dense than the salty seawater, the effluent moves upward and outward into the current as it disperses into an area called the mixing zone. At the same time, the current continues to move it away from the coastline. Natural processes eventually break down the effluent components, which are by now very highly diluted.

1.7.1 Mixing Zone

The mixing zone is very important for the dilution of wastewater in the sea. In the United States, the Environmental Protection Agency (US-EPA) regulations for toxics [6] define a mixing zone as:

An area where an effluent discharge undergoes initial dilution and is extended to cover the secondary mixing in the ambient water body. A mixing zone is an allocated impact zone where water quality criteria can be exceeded as long as acutely toxic conditions are prevented.

1.7.2 Deep Sea Discharge Outfalls

The major deep sea discharge outfalls for treated wastewater are summarized in Table 1.2.

Table 1.2 Deep sea discharge outfalls for treated wastewater discharge.

Location	Country	Distance from shore (km)
Honolulu (Honouliuli WWTP)	USA	2.67
Southern California Bight (Point Loma WWTP, San Diego)	USA	7.24
Santa Monica (Hyperion Water Reclamation Plant)	USA	8.1
Boston (Deer Island WWTP)	USA	15
Anglesea, Victoria (Barwon Water)	Australia	0.7
Geelong, Victoria (Black Rock Water Reclamation Point outfall)	Australia	1.2
Sydney (Malabar Island)	Australia	2.6
Cape Town (Green Point outfall)	South Africa	1.6
Cape Town (Camps Bay outfall)	South Africa	1.4
Cape Town (Hout Bay outfall)	South Africa	2.1
Ankleshwar, Gujarat (NCTL)	India	9.5
Mumbai (government sewage treatment plant, Bandra Reclamation)	India	3
Ipanema Beach, Rio de Janeiro	Brazil	4.3

NCTL, Narmada Clean Tech Ltd; WWTP, waste water treatment plant.

1.7.3 Wastewater Discharge Norms for Deep Sea Discharge Outfalls

The US-EPA has declared wastewater discharge norms for deep sea discharge outfalls (navigable water) based on effluent limitations guidelines representing the degree of effluent reduction attainable by application of the best practicable control technology currently available. India has declared general norms for wastewater discharge applicable for all types of discharge. These norms are summarized in Table 1.3.

Table 1.3 India vs US-EPA norms for deep sea discharges for pesticide and pharmaceutical industries.

Parameters	Pesticide industries effluent discharge[a] standards			Pharmaceutical industries effluent discharge standards	
	India[b]	US-EPA[c]		India[b]	US-EPA[d]
pH	6.0–9.0	6.0–9.0	6.0–9.0	6.0–9.0	6.0–9.0
COD	250 ppm	13 kg/ton production	4333 ppm	250 ppm	1675 ppm
BOD	100 ppm	7.4 kg/ton production	2466 ppm	100 ppm	267 ppm
TSS	100 ppm	6.1 kg/ton production	2033 ppm	100 ppm	472 ppm
Ammoniacal nitrogen	50 ppm	No limit	No limit	50 ppm	No limit
Oil and grease	10 ppm	NA	NA	10 ppm	NA

[a] For pesticide industries, average effluent discharge of $3 m^3$/ton of production is assumed.
[b] Indian Central Pollution Control Board (CPCB) standards: http://cpcb.nic.in/displaypdf.php?id=sw5kdxn0cnktu3bly2lmawmtu3rhbmrhcmrzl0vmzmx1zw50lzqznc0xlnbkzg.
[c] US-EPA CFR Part 455 Pesticide industry effluent standard: https://www.epa.gov/eg/pesticide-chemicals-effluent-guidelines.
[d] Pharma US-EPA: https://www.epa.gov/sites/production/files/2015-10/documents/pharmaceutical-permit-guidance_2006.pdf.

1.8 Environmental Rule of Law

To achieve sustainable development, an environmental rule of law is a must. Environmental rule of law ensures a fair society, living within environmental limits, and creating a sustainable future for all. It is also a barometer for the health of government institutions that avre held accountable by an informed and engaged public.

Environmental laws have grown dramatically since the early 1990s, as countries have come to understand the vital links between the environment, economic growth, public health, social cohesion, and security [7]. As of 2017, 176 countries have environmental framework laws; 150 countries have enshrined environmental protection or the right to a healthy environment in their constitutions; and 164 countries have created cabinet-level bodies responsible for environmental protection. These and other environmental laws, rights, and institutions have helped to slow – and in some cases to reverse – environmental degradation and to achieve the public health, economic, social, and human rights benefits that accompany environmental protection.

1.8.1 The Polluter Pays Principle

The "polluter pays" principle is the commonly accepted practice that those who produce pollution should bear the costs of managing it to prevent damage to human health and/or the environment. For instance, a factory that produces a potentially poisonous substance as a by-product of its activities is usually held responsible for its safe disposal. The polluter pays principle is part of a set of broader principles to guide sustainable development world-wide known as the 1992 Rio Declaration [8]. Principle 16 states:

> National authorities should endeavour to promote the internalization of environmental costs and the use of economic instruments, taking into account the approach that the polluter should, in principle, bear the cost of pollution, with due regard to the public interest and without distorting international trade and investment.

1.9 Trends in Wastewater Treatment Technology

In the past, industries preferred the conventional design of wastewater treatment to meet statutory norms, but the current trend is toward modular design of wastewater treatment to meet the recycling norms; in future, the trend will be toward extreme modular design of wastewater treatment, fully equipped with smart technology. Figure 1.12 shows the trends in industries.

Past trend Conventional design (to meet statutory norms)	Current trend Modular design (to meet recycling norms)	Future trend Extreme modular design (using smart technology)
• Biological treatment to just meet the statutory norms. • Technology: ASP. • Higher footprint. • RCC tanks. • Lower treatment efficiency. • Manual control. • Higher operating cost.	• Advance biological treatment to meet the recycling & reuse norms. • Technology: MBBR-UF-RO. • Medium footprint. • Bolted MS prefab tanks. • Higher treatment efficiency. • Automized control. • Medium operating cost.	• Smart technology based wastewater treatment. • Technology: MBR-FO-scaleban • Small footprint. • Containerized systems. • Higher treatment efficiency. • Remote control. • Lower operating cost.

Wastewater treatment technology trend is toward extreme modular design with smart technology.

Figure 1.12 Wastewater treatment technology trends in industries.

References

1 United Nations (2017). *The United Nations World Water Development Report. Wastewater: The Untapped Resource.* New York: UN.

2 United Nations (2018). *SDG 6 Synthesis Report 2018 on Water and Sanitation.* New York: UN.

3 BCC Publishing (2018). *Water and Wastewater Treatment Technologies: Global Markets. ENV008E*. Wellesley, MA: BCC Research.

4 National Research Council (1993). *Managing Wastewater in Coastal Urban Areas*. Washington, DC: National Academies Press.

5 Sydney Water (2007). *Sydney's Deep-Water Ocean Outfalls. Long-Term Environmental Performance*. St Leonards, NSW: Australian Water Association.

6 United States Environmental Protection Agency (1991). *CFR Part 455 Pesticide Industry Effluent Standard*. Washington, DC: US-EPA.

7 United Nations Environment Programme (2019). *Environmental Rule of Law First Global Report*. Nairobi: UNEP.

8 United Nations. 1992. *Report of the United Nations Conference on Environment and Development*. Rio Declaration on Environment and Development. UN, New York.

Further Reading

Bleninger,T., Jirka, G.H., and Roberts, P.J.W. 2011. Mixing Zone Regulations for Marine Outfall Systems. International Symposium on Outfall Systems, Mar del Plata, Argentina.

Chaubey, M. (2002). Treatment of industrial wastewater with solardetoxification technology. *Environmental Pollution Control Journal*, March–April 2002, 5 (3): 36–39.

Chaubey, M. (2016). Assessment of aerobic biological technologies for wastewater treatment of F.M.C.G. industries. *Water Digest* 2016: 30–36.

Chaubey, M. (2019). Best practices & design considerations for wastewater treatment with MBBR technology. *Official Journal of Indian Chemical Council*, September 2019 Edition: 16–19.

Chaubey, M. (2020). Techno-economic feasibility study of zero liquid discharge. *Official Journal of Indian Chemical Council*, March 2020 Edition: 26–30.

CPHEEO (2013). *Manual on Sewerage and Sewage Treatment Systems – 2013*. New Delhi: Central Public Health & Environmental Engineering Organization, Ministry of Urban Development.

Feitosa, R.C. (2016). *Ocean Outfalls as an Alternative to Minimizing Risks to Human and Environmental Health*. Brazil: Escola Nacional de Saúde Pública, Fundação Oswaldo Cruz.

Kaushika, N.D. and Chaubey, M. (2004). Laboratory investigations of photocatalytic detoxification for the prevention of biological fouling in reverse osmosis membrane. *International Journal, Research Journal of Chemistry and Environment* 8 (2): 15–20.

Metcalf & Eddy Inc (2014). *Wastewater Engineering: Treatment and Resource Recovery, 5*. New Delhi: Tata McGraw-Hill.

Telangana Pollution Control Board. 1996. India General Standards for Discharge of Environmental Pollutants. Part A: Effluents for Marine Coastal Areas. https://tspcb.cgg.gov.in/environment/general%20standards%20for%20discharge%20of%20environmental%20pollutants.pdf.

2

Wastewater Characteristics

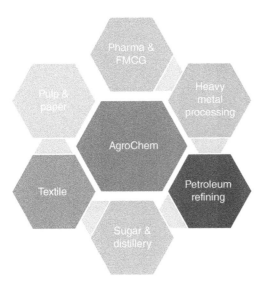

CHAPTER MENU

2.1 Wastewater Characteristics of Various Industries

Understanding wastewater characteristics is key for the design of any wastewater treatment plants. In this chapter, the wastewater characteristics of various industries based on the author's global experience of 23 years in various industries are summarized.

2.1.1 Agrochemical Industries Effluent Characteristics

The sources of effluent generated in agrochemical industries are from processing, cleaning, and washing of manufacturing equipment, production of condensate, boiler blowdown,

Wastewater Treatment Technologies: Design Considerations, First Edition. Mritunjay Chaubey.
© 2021 John Wiley & Sons Ltd. Published 2021 by John Wiley & Sons Ltd.

cooling tower blowdown, floor washing, and laboratory cleaning activities. Details of mixed raw effluent from technical units of agrochemical industries are summarized in Tables 2.1–2.3. The quantity of effluent generation is approx. 2–5 m^3/ton of production. However, effluent characteristics and generation quantity depend on the type of product manufactured.

Table 2.1 Effluent characteristics of technical units of agrochemical industries.

Parameters	Unit	Value
pH		1.5–12.5
Color	Pt-Co	500–700
Odor	–	Smelly
TSS	ppm	100–500
Temperature	°C	30–40
COD	ppm	5000–15 000
BOD	ppm	1500–5000
Ammonium group NH$_4$	ppm	50–200
Phenolic compounds	ppm	10–50
Sulfide	ppm	10–50
Cyanide	ppm	1–2
TKN	ppm	100–200
Free chlorine	ppm	0.5–1
Phosphate	ppm	5–50
TDS	ppm	5000–20 000
Calcium hardness as CaCO$_3$	ppm	100–2000
Magnesium hardness as CaCO$_3$	ppm	100–500
Silica	ppm	10–200
Sodium	ppm	2000–6000
P-alkalinity as CaCO$_3$	ppm	200–1000
Chloride	ppm	1000–5000
Sulfate as SO$_4{}^-$	ppm	3000–8000
Nitrate as NO$_3$	ppm	500–2500
Iron	ppm	0.5–2
Fluoride	ppm	0.5–2
Other heavy metal	ppm	0.5–2

Table 2.2 Low-strength effluent characteristics of technical units of agrochemical industries.

Parameters	Unit	Value
pH		1.5–12.5
TSS	ppm	100–200
COD	ppm	5000–10000
BOD	ppm	1500–4000
Ammonium group NH_4	ppm	50–200
Phenolic compounds	ppm	1–5
Sulfide	ppm	10–100
Cyanide	ppm	1–5
TKN	ppm	100–200
Phosphate	ppm	5–30
TDS	ppm	3000–10000
Calcium hardness as $CaCO_3$	ppm	100–2000
Magnesium hardness as $CaCO_3$	ppm	100–500
Silica	ppm	10–200
Sodium	ppm	1000–5000
Chloride	ppm	500–5000
Sulfate as SO_4^-	ppm	1000–5000
Nitrate as NO_3	ppm	100–300
Other heavy metal	ppm	0.5–2

Table 2.3 High-strength effluent characteristics of technical units of agrochemical industries.

Parameters	Unit	Value
pH		1.5–12.5
TSS	ppm	100–800
COD	ppm	10000–30000
BOD	ppm	5000–10000
Ammonium group NH_4	ppm	100–500
Phenolic compounds	ppm	10–50
Sulfide	ppm	100–1000
Cyanide	ppm	10–50
TKN	ppm	100–500
Phosphate	ppm	50–150
TDS	ppm	10000–30000
Calcium hardness as $CaCO_3$	ppm	500–6000
Magnesium hardness as $CaCO_3$	ppm	1000–3000

(*Continued*)

Table 2.3 (Continued)

Parameters	Unit	Value
Silica	ppm	50–200
Sodium	ppm	6000–15000
P-alkalinity as $CaCO_3$	ppm	200–1000
Chloride ·	ppm	5000–10000
Sulfate as SO_4^-	ppm	5000–15000
Nitrate as NO_3	ppm	500–1000
Iron	ppm	0.5–2
Fluoride	ppm	0.5–10
Other heavy metal	ppm	0.5–2

2.1.2 Textile Industries Effluent Characteristics

The textile industry is water intensive and produces pollutants of different forms. The freshwater required for processing textiles is approx. $200\,m^3$/ton of production. Major environmental issues result from wet processing, which includes sizing, desizing, scouring, bleaching, mercerization, dyeing, washing, and heat-setting activities. Details are summarized in Table 2.4. The quantity of effluent generation is approx. $50–120\,m^3$/ton of production. However, effluent characteristics and generation quantity depend on the type of textile product manufactured.

Table 2.4 Effluent characteristics of textile industries.

Parameters	Unit	Value
pH		5–12
COD	ppm	1500–4000
BOD	ppm	500–1500
TDS	ppm	3000–8000 (in some specific product production very high ~15000 ppm)
TSS	ppm	3000–8000 (in some specific product production very high ~15000 ppm)
Chloride	ppm	500–1000
Fluoride	ppm	0.5–2
O&G	ppm	2–10
Phenolic compounds	ppm	0.5–5
Cyanide	ppm	0.2–2
Anionic detergents	ppm	10–20

(Continued)

Table 2.4 (Continued)

Parameters	Unit	Value
Sulfate as SO_4	ppm	500–1500
Sulfide	ppm	800–1200
Ammonia	ppm	5–20
Total toxic metals	ppm	0.5–2
Zinc	ppm	0.2–1
Iron	ppm	0.5–2
Boron	ppm	0.3–1
Chlorine (residual)	ppm	0.5–1
TKN	ppm	10–30
Color	Pt-Co	200–700
Total phosphorus	ppm	5–20
Total hardness as $CaCO_3$	ppm	200–400
Calcium as $CaCO_3$	ppm	100–200
Magnesium as $CaCO_3$	ppm	100–200
Total silica	ppm	20–40

2.1.3 Pharmaceutical Industries Effluent Characteristics

Pharmaceutical effluent is very hard to degrade in view of the chemical and physical properties of the waste components. In addition, there are concerns expressed by some for the management of chemicals and conditions that can generate antibiotic-resistant microorganisms. Sources of effluent generation in pharmaceutical manufacturing industries comprise five categories: fermentation products, extraction products, chemical synthesis products, mixing/compounding and formulation, and research subcategories of the manufacturing of products for internal and external consumption such as tablets, capsules, ointments, powders, and solutions. Details of mixed raw effluent from pharmaceutical manufacturing industries are summarized in Tables 2.5–2.7. The quantity of effluent generation is approx. 3–5 m^3/ton of production. However, effluent characteristics and generation quantity depend on the type of pharmaceutical product manufactured.

Table 2.5 Effluent characteristics of pharmaceutical manufacturing industries.

Parameters	Unit	Value
pH		3–11
Color	Pt-Co	50–200
TSS	ppm	500–3000
Calcium hardness as $CaCO_3$	ppm	1000–2000
Magnesium hardness as $CaCO_3$	ppm	500–1000

(Continued)

Table 2.5 (Continued)

Parameters	Unit	Value
TDS	ppm	3500–10 000
BOD	ppm	3000–5000
COD	ppm	5000–12 000
Total alkalinity, $CaCO_3$	ppm	200–800
Chloride as Cl^-	ppm	1500–3000
Sulfate as SO_4^-	ppm	500–2500
TKN	ppm	50–200
Ammoniacal nitrogen as NH_3-N	ppm	50–150
O&G	ppm	10–200
Sulfide	ppm	1–2
Total silica	ppm	2–5
Iron	ppm	5–20
Zinc	ppm	5–10
Fluoride	ppm	2–5
Phosphate	ppm	0–10

Table 2.6 Low-strength effluent characteristics of pharmaceutical manufacturing industries.

Parameters	Unit	Value
pH	s.u	7.0–9.0
TSS	ppm	500–1000
TDS	ppm	3000–8000
BOD	ppm	3000–5000
COD	ppm	5000–10 000
Sulfate	ppm	50–200
Chloride	ppm	200–500
Calcium	ppm	200–400
Alkalinity	ppm	500–1000
Ammoniacal nitrogen	ppm	50–200

Table 2.7 High-strength effluent characteristics of pharmaceutical manufacturing industries.

Parameters	Unit	Value
pH	s.u	6.0–9.0
TSS	ppm	500–2000

(Continued)

Table 2.7   (Continued)

Parameters	Unit	Value
TDS	ppm	10 000–40 000
BOD	ppm	10 000–15 000
COD	ppm	15 000–45 000
Sulfate	ppm	1000–5000
Chloride	ppm	2000–15 000
Calcium	ppm	500–3500
Alkalinity	ppm	5000–7000
Ammoniacal nitrogen	ppm	200–1000

2.1.4   Heavy Metal Processing Industries Effluent Characteristics

The two main sources of heavy metals in wastewater are natural and human. Natural factors include soil erosion, volcanic activities, urban run-offs, and aerosol particulates, while human factors include metal finishing and electroplating processes, mining extraction operations, textile industries, and nuclear power. When this happens, toxic metals may be released into wastewater effluents. The most common toxic heavy metals in wastewater are arsenic, lead, mercury, cadmium, chromium, copper, nickel, silver, and zinc. The release of high amounts of heavy metals into water bodies creates serious health and environmental problems and leads to the requirement of wastewater treatment plants to meet standard discharge norms. The persistence of heavy metals in wastewater is due to their non-biodegradable and toxic nature. Sources are coke plants, blast furnaces, chemical by-products and chemical processes, water-cooled rolls, pumps, extrusion experiment, and transfer lines for sludges and slurries. This may be through either rinsing of the product or spillage and dumping of process baths. Details of mixed raw effluent from heavy metal processing industries are summarized in Table 2.8. The quantity of effluent generation is approx. 1–3 m^3/ton of production. However, effluent characteristics and generation quantity depend on the type of heavy metal processing industry.

Table 2.8   Effluent characteristics of heavy metal processing industries.

Parameters	Unit	Value
pH		4–10
Total hardness	ppm	500–1500
Chloride	ppm	500–2300
Sulfate	ppm	2000–5000 (in some specific product production very high ~10 000 ppm)
COD	ppm	300–500
BOD	ppm	50–100

(Continued)

Table 2.8 (Continued)

Parameters	Unit	Value
Silica	ppm	3–20
TDS	ppm	2000–10 000 (in some specific product production very high ~20 000 ppm)
Turbidity	NTU	5–20
TSS	ppm	50–200
Heavy metal	ppm	1–5
Dissolved iron	ppm	1–3
Zinc	ppm	1–5
O&G	ppm	1–5

2.1.5 Petroleum Refining Industries Effluent Characteristics

Petroleum refineries process raw crude oil into three categories of products:

- **Fuel products**: Fuel products include gasoline, distillate fuel oil, jet fuels, residual fuel oil, liquefied petroleum gases, refinery fuel, coke, and kerosene.
- **Non-fuel products**: Non-fuel products include asphalt and road oil, lubricants, naphtha solvents, waxes, non-fuel coke, and miscellaneous products.
- **Petrochemicals and petrochemical feedstocks**: Petrochemicals can be defined as a large group of chemicals derived from petroleum and natural gas and used for a variety of chemical purposes. Petrochemical products include naphtha, ethane, propane, butane, ethylene, propylene, butylene, and BTEX compounds (benzene, toluene, ethylbenzene, and xylene).

The major sources of effluent generation streams from petroleum refineries industries are shown in Table 2.9. The detailed characteristics of effluent from these industries are summarized below. The overall quantity of effluent generation from petroleum refineries process industries per ton of production is approx. 0.5–5 m³. According to general guidelines, for per ton of crude processing approx. 3.5–5 m³ of wastewater is generated when cooling water is recycled. However, the quantity of wastewater generated and its characteristics depend on the petroleum refinery process configuration and type of crude processing.

Table 2.9 Major wastewater streams.

Wastewater streams	Description
Desalter water	Water produced from washing the raw crude prior to topping operations.
Sour water	Wastewater from steam stripping and fractionating operations that comes into contact with the crude being processed.
Other process water	Wastewater from product washing, catalyst regeneration, and dehydrogenation reactions.

(Continued)

Table 2.9 (Continued)

Wastewater streams	Description
Spent caustic	Formed in extraction of acidic compounds from product streams.
Tank bottoms	Bottom sediment and water settles to the bottom of tanks used to store raw crude. The bottoms are periodically removed.
Cooling tower	Once-through cooling tower water and cooling tower blowdown to prevent buildup of dissolved solids in closed-loop cooling systems.
Condensate blowdown	Blowdown from boilers and steam generators to control buildup of dissolved solids.
Source water treatment system	Source water must be treated prior to use in the refinery. Waste streams may include water from sludge dewatering if lime softening is used; ion exchange regeneration water; or reverse osmosis wastewater.
Stormwater	Process area and non-process area runoff from storm events.
Ballast water	Ballast water from product tankers and ships.

2.1.5.1 Effluent Characteristics from Oil Refining Operations

The **low total dissolved solids (LTDS) stream** is a mixture of process and oily waters which include non-phenolic wastewaters, treated to an effluent quality adequate for reuse for cooling water makeup, fire water makeup, and irrigation water for development and maintenance of the local green belt.

The **high total dissolved solids (HTDS) stream** is a mixture of process and oily waste-waters that have been in contact with process streams, such as in crude unit desalters, and have absorbed or dissolved mineral ions such as sodium chloride. This stream also contains (treated neutralized) process solvents such as spent caustics and phenolic wastewater. This water is treated to an effluent quality adequate for reuse as partial makeup water in cooling tower. (See Tables 2.10–2.13.)

Table 2.10 Effluent characteristics of oil refining operations.

Parameters	Unit	Value
pH		6–10
TDS	ppm	Low streams 500–1000 High streams 1000–4000
TSS	ppm	200–500
Turbidity	NTU	300–500
Color	Pt-Co	2000–5000
BOD	ppm	200–500
COD	ppm	500–2000

(Continued)

Table 2.10 (Continued)

Parameters	Unit	Value
Ammoniacal nitrogen	ppm	20–50
O&G	ppm	50–1000 (in desalter from tank bottom ~5000 ppm)
Phenols	ppm	20–200
Sulfide	ppm	10–100
Sulfate	ppm	50–500
TKN	ppm	50–100
Heavy metal	ppm	1–20
Benzene	ppm	1–100
Chromium	ppm	0.1–100
Lead	ppm	0.2–10

Table 2.11 Effluent characteristics of petrochemical manufacturing industries.

Parameters	Unit	Value
pH		6–7.5
O&G	ppm	500–8000 (free oil and emulsified oil)
BOD	ppm	500–2000
COD	ppm	2000–6000
TSS	ppm	200–500
Phenols	ppm	10–50
Sulfide	ppm	10–20
Cyanide	ppm	1–5
Heavy metal	ppm	1–3
TDS	ppm	1000–4000
Ammonia nitrogen	ppm	20–50
TKN	ppm	50–100
Total hardness as $CaCO_3$	ppm	500–1000
Magnesium hardness as $CaCO_3$	ppm	250–300
Calcium hardness as $CaCO_3$	ppm	500–800
Total silica	ppm	20–50
Dissolved iron as Fe	ppm	1–3
Chloride	ppm	50–200

Table 2.12 Effluent characteristics of spent caustic effluent streams.

Parameters	Unit	Value
pH		10–13
Total sulfides	ppm	10 000 (in the form of $Na_2S/NaSH$)
Hydrocarbons	ppm	500
COD	ppm	25 000
NaOH	%	0.05–0.1

Table 2.13 Effluent characteristics of gas well exploration operations.

Parameters	Unit	Value
pH		5–9
Temperature	°C	4–70
Total oil content	ppm	10 000–15 000
TSS	ppm	50–200
COD	ppm	10 000–20 000
BOD	ppm	4000–10 000
TDS	ppm	1000–2500

2.1.6 Pulp and Paper Industries Effluent Characteristics

Pulp and paper industries are comparatively more polluting industries and highly water intensive. Average freshwater consumption per ton of paper is approx. $150 m^3$. However, freshwater consumption depends on the type of raw material used in the paper mill, whether wood based, agro based, or waste-paper based. The major sources of wastewater generation in an integrated pulp and paper industry are as follows:

- **Raw material section**: Washing of wooden chips in large-scale pulp and paper mills using wool as raw material, washing of bagasse for separation of pith, and washing of rice or wheat before pulping.
- **Pulping and bleaching**: Washing of chemically cooked pulp, washing of pulp during bleaching, and washing of pulp cleaning equipment.
- **Stock preparation and paper machine**: Cleaning of pulp in cleaning equipment, filtration for wire section of paper machine and paper machine presses.
- **Chemical recovery**: Foul condensate from evaporator and steam surface condenser and boiler blowdown.

Besides the above major sources of wastewater generation there are frequent leakages of black liquor from pump glands and its improper handling, which contribute significant color and pollution to the stream. Color of effluent from paper machines is blackish in nature. The wastewater generated has the following impurities:

- Suspended solids including bark particles, fiber, pigments, and dirt.
- Dissolved colloidal organics like hemicelluloses, sugars, lignin compounds, alcohols, turpentine, sizing agents, and adhesives such as starch and synthetics.
- Color bodies, primarily lignin compounds and dyes.
- Dissolved inorganics such as NaOH, Na_2SO_4, and bleach chemicals.
- Thermal loads.
- Toxic chemicals.

The quantity of effluent generation from pulp and paper industries per ton of paper production is approx. $50-75\,m^3$. However, effluent characteristics and generation quantity depend on the type of pulp and paper industries operation and type of raw material used (see Table 2.14).

Table 2.14 Effluent characteristics of pulp and paper industries.

Parameters	Unit	Value
pH		4.0–12.0
Temperature	°C	40–70
Suspended solids	ppm	500–3000
BOD	ppm	1000–4000
COD	ppm	3000–8000
Ammoniacal nitrogen	ppm	20–100
Alkalinity	ppm	200–1000

2.1.7 Sugar and Distillery Industries Effluent Characteristics

Sugar industries have an important place in Indian economic development. However, wastewater generated from these industries has a high degree of pollution load. Sugar industries in India generate approx. $1-2\,m^3$ of wastewater per ton of sugar cane crushed.

Sources of wastewater in the production of alcohols and alcoholic beverages from distillery industries are from wash water, steam water, cleaning and cooling waters, and stillages. Digested molasses spent wash is a highly recalcitrant waste product, which does not decompose by the usual biological treatments. Spent wash is the most complex and strongest organic effluent in distillery industries. It has a very high pollution load with a very high chemical oxygen demand (COD) and biological oxygen demand (BOD). The main components of wastewater are carbohydrates, proteins, amino acids, and glycerol that accounts for a major portion of COD. The quantity of effluent generation from distillery industries per ton of alcohol production is approx. 10–15 m^3. However, effluent characteristics and generation quantity depend on the type of distillery industry and raw material used (see Tables 2.15 and 2.16).

Table 2.15 Effluent characteristics of sugar industries.

Parameters	Unit	Value
pH		4.0–8.0
Temperature	°C	50–80
TSS	ppm	2000–15 000
COD	ppm	75 000–6 500 000
BOD	ppm	30 000–70 000
TDS	ppm	70 000–7 500 000
Total volatile solids	ppm	40 000–70 000
TKN	ppm	500–2000
Potash as K	ppm	10 000–2 500 000
Phosphate as PO$_4$	ppm	1500–6000
Sodium	ppm	100–500
Chloride	ppm	4000–10 000
Sulfate as SO$_4$	ppm	1500–8000
O&G	ppm	10–20

Table 2.16 Effluent characteristics of distillery industries.

Parameters	Unit	Value
pH	s. u.	3.0–8.5
Temperature	°C	30–55
TDS	ppm	1500–3000
TSS	ppm	500–1000
COD	ppm	5000–10 000
BOD	ppm	5000–8000
O&G	ppm	20–50
Volatile acids	ppm	3000–3500
TKN	ppm	100–500
Potash as K	ppm	500–1000
Sodium	ppm	100–500
Chloride	ppm	1000–5000
Sulfate as SO$_4$	ppm	500–3000

2.1.8 Dairy Industries Effluent Characteristics

Water plays a key role in milk processing. It is used in every step of the technological line, including cleaning and washing, disinfection, heating, and cooling. Dairy industries are water intensive. Wastewater is generated during the transformation of raw milk into pasteurized and sour milk, yogurt, hard and soft cheeses, cottage cheese, cream, butter products, ice cream, milk and whey powders, lactose, and condensed milk, as well as various types of desserts. Due to the rapid industrialization of milk production, dairy processing is usually considered the largest industrial food wastewater source. Dairy effluents are distinguished by their relatively increased temperature, high organic content, and wide pH range, which requires special treatment to meet standard discharge norms and to reduce environmental impact.

The quantity of effluent generation from dairy industries per ton of milk processed is approx. 1–3 m^3. However, effluent characteristics and generation quantity depend on the type of operation and type of product production.

Major fluctuations in wastewater quality and quantity are very problematic because each milk product needs a separate technological line. This results in a change of dairy effluent composition with the start of a new cycle in the manufacturing process. Seasonal variations also contribute to a higher dairy plant load in summer than in winter. On average, wastewater discharge is 70% of the amount of the freshwater used at the plant. Also, milk loss in wastewater is around 0.5–4% of the processed milk (see Table 2.17).

Table 2.17 Effluent characteristics of dairy industries.

Parameters	Unit	Value
pH		4–12
BOD	ppm	1000–6000
COD	ppm	2000–10000
Fat, oil, and grease	ppm	50–1000
TSS	ppm	100–5000
TKN	ppm	50–150
Total phosphorus	ppm	15–60
Alkalinity carbonate	ppm	500–1200
TDS	ppm	1000–3000
Ammonia	ppm	2–6
Odor		Light smell
Color		Whitish

2.1.9 Fast Moving Consumer Goods Industries Effluent Characteristics

Wastewater generated in the production of fast moving consumer goods (FMCG) like processed foods, prepared meals, beverages, baked goods, frozen and dry goods, cleaning products, some medicines, cosmetics and toiletries, and office supplies from manufacturing industries are summarized below. However, effluent characteristics depend on the type of product manufacturing and operation (see Tables 2.18–2.24). Normally, wastewater generated from FMCG is in the range of 2–5 m³/ton of production.

Table 2.18 Wastewater characteristics of ice cream manufacturing.

Parameters	Unit	Value
TSS	ppm	1000–2000
pH	–	3–10
BOD	ppm	3000–5000
COD	ppm	8000–10000
O&G	ppm	1000–1500
TDS	ppm	1500–3000
Total phosphorus	ppm	20–200
TKN	ppm	35

Table 2.19 Wastewater characteristics of home care industries.

Parameters	Unit	Value
TSS	ppm	200–500
pH	–	6–8
BOD	ppm	2500–3500
COD	ppm	7000–10000
O&G	ppm	200–400
TDS	ppm	1500–3000

Total detergent is a key parameter for the design, to be defined case specific

Table 2.20 Wastewater characteristics of personal care industries.

Parameters	Unit	Value
TSS	ppm	1000–2000
pH	–	7–9
BOD	ppm	4000–7000
COD	ppm	8000–15000
O&G	ppm	200–400
TDS	ppm	1500–3000

Total detergent is a key parameter for the design, to be defined case specific

Table 2.21 Wastewater characteristics of spreads and dressings industries.

Parameters	Unit	Value
TSS	ppm	200–500
pH	–	6–9
BOD	ppm	2000–3500
COD	ppm	4000–7000
O&G	ppm	50–100
TDS	ppm	1500–3000

Table 2.22 Wastewater characteristics of savory fooods industries.

Parameters	Unit	Value
TSS	ppm	800–1400
pH	–	4–6
BOD	ppm	7500–8000
COD	ppm	15000–20000
O&G	ppm	100–1500
TDS	ppm	2000–2800

Table 2.23 Wastewater characteristics of oral (mouth and teeth cleaning) industries.

Parameters	Unit	Value
TSS	ppm	700–1400
pH	–	7–9
BOD	ppm	3000–6000
COD	ppm	7000–12000
O&G	ppm	150–300
TDS	ppm	1500–3000

Table 2.24 Wastewater characteristics of deodorant industries.

Parameters	Unit	Value
TSS	ppm	500–800
pH	–	5–7
BOD	ppm	1000–2000
COD	ppm	2500–3500
O&G	ppm	50–100
TDS	ppm	1500–3000

2.1.10 Fruit-Based Products Processing Industries Effluent Characteristics

Wastewater generated in the production of fruit-based products like palm oil, juices and concentrates, pulp, soft drinks, canned and dehydrated products, jams and jellies, pickles and chutneys has effluent characteristics as summarized in Table 2.25. However, these characteristics depend on the type of product manufacturing and operation.

Table 2.25 Effluent characteristics of fruit-based products processing industries.

Parameters	Unit	Value
pH		4.0–8.0
TSS	ppm	200–1000 (after primary filtration)
Fat, oil, and grease	ppm	200–500
BOD	ppm	5000–20 000
COD	ppm	10 000–40 000
TDS	ppm	1500–5000

2.1.11 Fertilizer Industries Effluent Characteristics

Wastewater generated in the production of fertilizer-based products like ammonia, nitrogen fertilizer, phosphate fertilizer, and potassium fertilizer has effluent characteristics as summarized in Table 2.26. The quantity of effluent generated from fertilizer industries per ton of product (ammonia/urea) is approx. 1–2.5 m^3. However, these characteristics and generation quantity depend on the type of fertilizer product, raw material used, manufacturing, and operation of plant.

Table 2.26 Effluent characteristics of fertilizer industries.

Parameters	Unit	Value
pH		6–9.5
TDS	ppm	1000–4000
COD	ppm	2000–10 000 (in some effluent streams ~50 000 ppm)
TSS	ppm	50–500
Ammoniacal nitrogen as NH_3-N	ppm	200–1500
Phosphate as PO_4	ppm	10–150
Cyanide	ppm	0.5–3
Alcohol	%	2–5 (in some streams)

2.1.12 Paint Manufacturing Industries Effluent Characteristics

Wastewater generated in the production of paints and coatings-based products has effluent characteristics as summarized in Table 2.27. However, these characteristics depend on the type of raw material used, paint product manufacturing, and operation of plant.

Table 2.27 Effluent characteristics of paint manufacturing industries.

Parameters	Unit	Value
pH		6–8
TDS	ppm	500–4000
TSS	ppm	4000–13 000
COD	ppm	5000–25 000
BOD	ppm	800–2500
TKN	ppm	50–500
Total phosphorus	ppm	1–15
Color	Pt-Co	20–50
O&G	ppm	50–300
Chloride	ppm	150–400
Heavy metal	ppm	1–10

2.1.13 Cement Plant Effluent Characteristics

Wastewater generated from cement plants has effluent characteristics that are inorganic in nature and very low organic impurities as summarized in Table 2.28.

Table 2.28 Effluent characteristics of cement plants.

Parameters	Unit	Value
pH		6–9
Temperature	°C	30–40
Turbidity	NTU	10–20
TDS	ppm	500–1000
Total alkalinity as $CaCO_3$	ppm	150–650
Total hardness as $CaCO_3$	ppm	250–800
Chloride	ppm	150–550
BOD	ppm	5–10
COD	ppm	10–50
TSS	ppm	300–1500 (physical particle also found)

2.1.14 Thermal Power Plant (Cooling Tower Blowdown) Effluent Characteristics

Wastewater generated from thermal power plant (cooling tower blowdown) industries has effluent characteristics as summarized in Table 2.29. However, cooling tower blowdown wastewater characteristics depend on the number of cycle of concentration (COC) of cooling water and cooling tower operation.

Table 2.29 Effluent characteristics of thermal power plants (cooling tower blowdown).

Parameters	Unit	Value
pH		6–9
TDS	ppm	800–1500
Turbidity	NTU	10–50
M-alkalinity as $CaCO_3$	ppm	50–200
Total hardness as $CaCO_3$	ppm	200–500
Calcium	ppm	50–200
Magnesium	ppm	50–150
Chloride	ppm	50–200
Sulfate	ppm	50–100
Sodium	ppm	20–100
Reactive silica	ppm	20–50
BOD	ppm	5–10
COD	ppm	20–50
TSS	ppm	10–50
Free residual chlorine	ppm	0.5

2.1.15 Smelter Plant (Aluminum Manufacturing) Effluent Characteristics

To produce aluminum products from bauxite, the following four processes are necessary:

- Mining: To extract raw bauxite from the earth.
- Refining: To refine raw bauxite into alumina powder.
- Smelting: To convert alumina into liquid aluminum.
- Casting: To produce Ingots, wire-rods, and billets from liquid aluminum.

Wastewater generated in aluminum smelter plants is in the range of 0.5–1 m^3/ton of aluminum produced. The characteristics of smelter process effluent are summarized in Table 2.30.

Table 2.30 Effluent characteristics of the smelter process.

Parameters	Unit	Raw effluent value
pH		6.4–7.9
Alkalinity	mg/l	55–170
Fluoride	mg/l	5–30
Total hardness	mg/l	68–164
Total iron	mg/l	0.015–1.27
Chloride	mg/l	16–72
Sulfate as SO_4^-	mg/l	15.1–42.75

(Continued)

Table 2.30 (Continued)

Parameters	Unit	Raw effluent value
Manganese	mg/l	0.1–1.15
TDS	mg/l	165–295
TSS	mg/l	33–360
COD	mg/l	10–40
BOD	mg/l	2–13.5
O&G	mg/l	1–28
Dissolved silica	mg/l	3.6–4.1
Colloidal silica	mg/l	BDL

2.1.16 Domestic Sewerage Wastewater Characteristics

In industrial operations, normally domestic sewerage wastewater generation is 50 l/capita/day. The characteristics of domestic sewerage are summarized in Table 2.31.

Table 2.31 Characteristics of raw domestic sewage.

Parameters	Unit	Value
TSS	ppm	100–150
pH		7–8
BOD	ppm	200–250
COD	ppm	350–450
O&G	ppm	15–25
TDS	ppm	100–1000

2.2 Wastewater Characteristics and Measuring Methodology

For daily operation and monitoring purposes, we need to measure various wastewater characteristics. Laboratory testing procedures for the following wastewater parameters are described below.

- pH value
- Total dissolved solids (TDS)
- Total suspended solids (TSS)
- Oil and grease
- Dissolved oxygen
- COD
- BOD
- Total Kjeldahl nitrogen (TKN)
- Total phosphorus

- Mixed liquor suspended solids (MLSS)
- Sludge volume index (SVI)
- Turbidity
- Color
- Temperature
- Acidity
- Alkalinity
- Hardness
- Iron
- Silica
- Sulfate
- Chloride
- Residual chlorine

2.2.1 pH Value

The term pH refers to the measurement of hydrogen ion activity in the solution. As a chemical component of wastewater, pH has a direct influence on wastewater treatability, regardless of whether treatment is physical, chemical, or biological. Water is composed of a positively charged hydrogen ion (H^+) and a negatively charged hydroxide ion (OH^-). In acidic (pH < 7) wastewater there is a high concentration of positive hydrogen ions, while in neutral wastewater the concentration of hydrogen and hydroxide ions is balanced. Basic (pH > 7) wastewater contains an excess of negative hydroxide ions. pH adjustment by addition of acidic (hydrochloric acid (HCl), H_2SO_4, etc.) or basic (NaOH, lime soda, etc.) chemicals is an important part of any wastewater treatment system. For better performance of biological treatment plants the pH required is in the range of 6.5–8.0.

Measurement of pH in wastewater treatment requires a quick, accurate, and robust method. Since the direct measurement of pH is very difficult, specific electrodes are needed for quick and accurate pH determination. pH in wastewater analysis is by online or lab method. For online monitoring a pH analyzer can be used with or without transmitter facility. The lab method is described below.

2.2.1.1 Preparation of Buffer Solutions

- pH 4 buffer solution: Dissolve 1.012 g anhydrous potassium hydrogen phthalate ($KHC_8H_4O_4$) in distilled water and make up to 100 ml in a volumetric flask.
- pH 7 buffer solution: Dissolve 1.361 g anhydrous potassium dihydrogen phosphate (KH_2PO_4) and 1.420 g anhydrous disodium hydrogen phosphate (Na_2HPO_4) (both of which have been dried at 110–130 °C for two hours) in distilled water and make up to 1000 ml in a volumetric flask.
- pH 9 buffer solution: Dissolve 3.81 g of sodium borate decahydrate (borax) ($Na_2B_4O_7$ $10H_2O$) in distilled water and make up to 1000 ml.

2.2.1.2 Standardization of pH Meter

- Wash the electrode thoroughly with distilled water.
- Dip the electrode in pH 7 buffer solution and set the reading at 7.

- Dip the electrode in pH 9 buffer solution and set the reading at 9.
- Again dip the electrode in pH 7 buffer solution and set the reading at 7.
- Dip the electrode in pH 4 buffer solution and set the reading at 4.

Before changing the solutions, rinse the electrodes with distilled water. Now the pH meter is standardized for taking pH of any sample.

2.2.2 Total Dissolved Solids

Total dissolved solids (TDS) is a measure of the dissolved combined content of all inorganic and organic substances that are present in water. Inorganic impurities like Na, Ca, Mg, Si, and Fe may be present in wastewater. If the TDS value of wastewater is <5000 ppm it can be easily treated in a wastewater treatment plant. If the TDS value is >5000 ppm it creates a problem in the plant, especially in the biological section. For proper design and operation of the plant, high TDS wastewater streams need to be identified and segregated.

TDS in wastewater can be analysed by online or lab method. For online TDS monitoring a conductivity analyzer can be used with or without transmitter facility.

TDS analysis by lab method is described as below.

2.2.2.1.1 Apparatus

- Evaporating dishes: Dishes of 100-ml capacity made of one of the following materials:
  - Porcelain 3 mm diameter
  - Platinum – generally satisfactory for all purposes
  - High silica glass.

- Desiccator provided with a desiccant containing a color indicator of moisture concentration.
- Glass fiber filter disks without organic binder.
- Filtration apparatus: One of the following, suitable for filter disks selected:
  - Membrane filter funnel
  - Gooch crucible 25–40 ml capacity, with Gooch crucible adapter
  - Filtration apparatus with reservoir and coarse (40–65 μm) fitted disk as filter support
  - Suction flask of sufficient capacity for sample size selected
  - Drying oven for operation at 180 °C
  - Steam-bath
  - Muffle furnace for operation at 550 °C.

2.2.2.1.2 Procedure

- **Preparation of glass fiber filter disk**: Insert disk with wrinkled side up into filtration apparatus. Apply vacuum and wash disk with three successive 20 ml volume of distilled water. Continue suction to remove all traces of water. Discard washings.
- **Preparation of evaporating disk**: Heat clean dish to 180 °C for one hour in an oven. Store in desiccator until needed. Weigh immediately before use.
- **Selection of filter and sample sizes**: Choose sample volume to yield between 2.5 and 200 mg dried residue. If more than 10 minutes are required to complete

filtration, increase filter size or decrease sample volume but do not produce less than 2.5 mg residue.
- **Sample analysis**: Filter measured volume of well-mixed sample to glass fiber filter, wash with three successive 10 ml volume of distilled water, allowing complete drainage between washing, and continue suction for about three minutes after filtration is complete. Transfer filtrate to a weighed evaporating dish and evaporate to dryness in a steam-bath. If filtrate volume exceeds dish capacity add successive portions to the same dish after evaporation. Dry for at least one hour in an oven at 180 °C, cool in desiccator to balance temperature, then weigh it. Repeat drying cycle of drying, cooling, desiccating, and weighing until a constant weight is obtained or unless weight loss is less than 4% of previous weight.

2.2.2.1.3 Calculation

$$Total\ Dissolved\ Solids(mg\,/\,l) = \left[(A - B) \times 1000\right] / Sample\ Volume(ml)$$

where
> A = Weight of dried residue + dish in mg
> B = Weight of dish in mg

2.2.3 Total Suspended Solids

Total suspended solids (TSS) value indicates TSS impurities that are present in wastewater. TSS include large floating materials and smaller physical materials such as silt, etc. To remove large floating materials various sizes of screen are installed, and for removal of smaller physical impurities coagulation and flocculation processes are used. The pretreatment section of the wastewater treatment plant should be properly designed based on inlet TSS value and considering other parameters, such as turbidity, oil and grease (O&G).

TSS in wastewater analysis is by online or lab method. For online TSS monitoring a TSS analyzer can be used with or without transmitter facility. TSS analysis by lab method is described below.

2.2.3.1.1 Apparatus
Gooch crucible (G3), filtration flank, rotary vacuum pump, asbestos powder.

2.2.3.1.2 Procedure
- Prepare Gooch crucible with G3 sintered disc by forming a layer of asbestos on it. For this, prepare asbestos solution by dissolving asbestos powder in distilled water and mixing it thoroughly, then allowing asbestos to settle. Take the supernatant and pour into a Gooch crucible and apply vacuum on the other side through a filtration flask; a layer of asbestos will be formed on the sintered disc. Dry the Gooch crucible in an oven at 105 °C and then cool it in a desiccator and weigh it (W1; see 'Calculation' below).
- Fit the Gooch crucible on a filtration flask and connect the filtration flask to the vacuum pump. Take a 25-ml sample, diluted to 200 ml, and pour it into the Gooch crucible slowly, then apply a vacuum.

- When the sample has passed the Gooch crucible, fill it with 20–25 ml distilled water and apply vacuum, so that all the distilled water passes through the crucible, no moisture is left in it and only TSS is retained.
- Dry the Gooch crucible with TSS in an oven at 105 °C for 30–60 minutes, then cool it in a desiccator and weigh it (W2).
- After completion of the Gooch test, clean the crucible by keeping 20 ml of chromic acid in it overnight and in the morning clean it and prepare again with asbestos solution.

2.2.3.1.3 Calculation

$$TSS\left(mg\,/\,l\right)=\left[\left(W2-W1\right)\times 1000\right]/\,Volume\ of\ sample\ in\,ml$$

where

W1 = Weight of filter + dish in mg
W2 = Weight of filter and dish + residue in mg

2.2.4   Oil and Grease

Oil and grease (O&G) value indicates oily impurities that are present in wastewater in the form of free and emulsified. These form a layer on top of the wastewater that affects the wastewater treatment plant, especially the biological treatment process. So there is a need to remove O&G in the pretreatment process, and various types of removal system can be considered including American plate interceptor (API), tilted plate interceptor (TPI), and dissolved air floatation (DAF) with proper dosing system.

O&G analysis is by online or lab method. For online monitoring an analyzer can be used with or without transmitter facility. The lab method is described below.

2.2.4.1.1 Reagents

- Magnesium sulfate solution: Dissolve 1 g of magnesium sulfate heptahydrate in 100 ml of water.
- Milk of lime: Mix 2 g of calcium oxide with water into a paste and dilute the suspension to 100 ml.
- Light petroleum (petroleum ether): Boiling range 40–60 °C.
- Dilute HCl.
- Anhydrous sodium sulfate.

2.2.4.1.2 Procedure

Take 250 ml, or an aliquot containing 50–150 mg, of extractable matter of the well-mixed sample in a beaker. If a noticeable layer of floating matter is present, carefully transfer as much of it as possible by decantation into a separating funnel. Draw into the beaker containing the residual portion of the sample any liquid that separates out in the funnel. To the sample in the beaker add 5 ml magnesium sulfate solution. Stir in a rotatory direction with a glass rod and continuously add small amounts of milk of lime until flocculation occurs. Continue stirring for two minutes, withdraw the glass rod, and wash it down in the separating funnel with a small quantity of light petroleum. Allow the precipitates in the beaker to settle for five minutes. When settled completely, siphon off the clear supernatant liquid to within about 1 cm of the top of the sediment. Allow any remaining floating oil to be in the beaker itself. Dissolve the precipitate in the beaker with dilute HCl and transfer the contents to a separating funnel, taking care not to transfer any large adventitious solids like twigs, leaves, etc. Rinse the beaker with about 50 ml of light petroleum and add this to the liquid in the funnel. Shake the funnel continuously, but not vigorously, for one minute. Allow the liquid layers to separate. Draw the aqueous layer into another separating funnel and extract again with a fresh 50-ml portion of light petroleum. Reject the aqueous layer and combine the petroleum extracts.

Add to the combined petroleum extracts 2 g of powdered anhydrous sodium sulfate and shake intermittently over a period of about 30 minutes. Filter through a small-size filter paper (Whatman No. 30), collecting the filtrate in a dry weighed wide-neck glass of 250 ml capacity. Wash the paper with two successive 20-ml portions of light petroleum and collect the filtrate in the flask. Distil off most of the light petroleum from the filtrate in the flask and, finally, evaporate the last traces in a current of warm air. Keep in a water-bath for 10 minutes, wipe the outside dry with a filter paper, cool in a desiccator, and weigh. The difference in weight is the weight of the residue. If after the solvent has evaporated the residue contains water, add 2 ml of acetone and evaporate in a water-bath. Repeat the acetone addition and evaporation until the contents are free of water.

Note: Some effluents do not readily flocculate with lime. In such cases, determine the suitable flocculating agent by preliminary trial and add them. The following flocculating agents are suggested:

- Aluminum sulfate – 1% solution with pH adjustment of the sample.
- Ferric chloride – 1% solution and ammonium hydroxide.
- Zinc acetate – 10% solution and sodium carbonate 5% solution.

2.2.4.1.3 Calculation

$$O \& G(mg / l) = W \times 1000 / V$$

where

W = Weight of residue in mg
V = Volume of sample taken for test in ml

2.2.5 Dissolved Oxygen

Dissolved oxygen (DO) is oxygen present in water. DO value is measured in the biological section of wastewater treatment plants. It is added to the aeration basin to enhance the oxidation process by providing oxygen to aerobic microorganisms so they can successfully turn organic wastes into inorganic byproducts. DO in aeration tanks should be maintained in the range of 2–3 ppm and it is required for proper functioning of the biological section.

If the DO is less than 2 ppm, the microorganisms in the center of the floc may die since the microorganisms on the outside of the floc use up the DO first. If this happens, the floc breaks up which means biological performance gets drastically reduced. If the DO content is too low, the environment is not stable for these microorganisms and they will die due to anerobic zones, the sludge will not be properly treated, and a time-consuming biomass replacement process will be needed. When the DO levels become too high, energy is wasted, expensive aeration equipment undergoes unnecessary usage, and unwanted organisms like filamentous microorganisms will grow in the aeration system.

DO in wastewater analysis is by online or lab method. For online DO monitoring an analyzer can be used with or without transmitter facility. An aeration basin with online DO measurement system can be installed to maintain the correct amount of DO. DO analysis by lab method is described below using the azide modified method.

2.2.5.1.1 Reagents

- Manganous sulfate solution: Dissolve 480 g $MnSO_4.4H_2O$, 400 g $MnSO_4.2H_2O$, or 364 g $MnSO_4.H_2O$ in distilled water, filter, and dilute to 1 l. The $MnSO_4$ solution should not give a color with starch when added to an acidified potassium iodide (KI) solution.
- Alkali iodide azide reagent:
  - For saturated or less than saturated samples: Dissolve 500 g NaOH and 135 g NaI in distilled water and dilute to 1 l. Add 10 g NaN_3 dissolved in 40 ml distilled water. Potassium and sodium salts may be used interchangeably. This reagent should not give a color with starch solution when diluted and acidified.
  - For supersaturated samples: Dissolve 10 g NaN_3 in 500 ml distilled water. Add 480 g NaOH and 750 g sodium iodide and stir until dissolved. There will be a white turbidity due to sodium carbonate, but this will do no harm. Do not acidify this solution because toxic hydrazoic acid fumes may be produced.
- Sulfuric acid concentration: 1 ml is equivalent to about 3 ml alkali iodide azide reagent.
- Starch solution: Use an aqueous solution of soluble starch powder mixtures. To prepare, dissolve 2 g of laboratory grade soluble starch and 0.2 g salicylic acid, as a preservative, in 100 ml hot distilled water.
- Standard sodium thiosulfate reagent: Dissolve 6.205 g of $Na_2S_2O_3.5H_2O$ in distilled water. Add 0.4 g solid NaOH and dilute to 1000 ml. Standardize with bi-iodate solution.
- Standard potassium bi-iodate solution 0.0021 M: Dissolve 812.4 mg $KH(IO_3)_2$ in distilled water and dilute to 1000 ml.
  - **Standardization**: Dissolve approximately 2 g KI, free from iodate, in an Erlenmeyer flask with 100–150 ml distilled water. Add 1 ml 6 N H_2SO_4 or a few drops of H_2SO_4 and 20 ml standard bi-iodate solution. Dilute to 200 ml and titrate liberated iodine with thiosulfate titrant, adding starch toward the end of titration, when a pale straw color is reached. When the solutions are of equal strength, 20 ml of 0.0021 M $Na_2S_2O_3$ should be required; if not, adjust the $Na_2S_2O_3$ solution to 0.0021 M.
- Potassium fluoride solution: Dissolve 40 g $KF.2H_2O$ in distilled water and dilute to 100 ml.

2.2.5.1.2 Procedure

- To the sample collected in a 250- to 300-ml bottle, add 1 ml $MnSO_4$ solution followed by 1 ml alkali iodide azide reagent. If pipettes are dipped into the sample, rinse them before returning them to reagent bottles. Alternatively, hold pipette tips just above the liquid surface when adding reagents. Stopper carefully to exclude air bubbles and mix by inverting bottle a few times. When precipitate has settled sufficiently (to approximately half the bottle volume) to leave clear supernatant above the manganese hydroxide floc, add

1.0 ml concentrated (conc.) H_2SO_4. Restopper and mix by inverting several times until dissolution is complete. Titrate a volume corresponding to 200 ml of the original sample after correction for a sample loss by displacement with reagents. Thus, for a total of 2 ml (1 ml each) of $MnSO_4$ and alkali iodide azide reagent in a 300-ml bottle, titrate $(200 \times 300)/(300-2) = 201$ ml.

- Titrate with 0.0021 M $Na_2S_2O_3$ solution of pale straw color. Add a few drops of starch solution and continue titration to first disappearance of blue color. If end point is over-run, back titrate with 0.0021 M bi-iodate solution added dropwise, or by adding a measured volume of treated sample, then correct for amount of bi-iodate solution or sample. Disregard subsequent recolorations due to the catalytic effect of nitrite or to traces of ferric salts that have not been complexed with fluoride.

2.2.5.1.3 Calculation

$$DO\left(mg\,/\,l\right) = \left(Burette\ reading,\ ml \times N\ of\ Na2S2O3 \times 8 \times 1000\right)/\ Volume\ of\ sample,\ ml$$

2.2.6 Chemical Oxygen Demand

Chemical oxygen demand (COD) value is a measurement of the oxygen required to oxidize soluble and particulate organic matter in water. If the COD value of wastewater is <10 000 it can be easily treated in a wastewater treatment plant in a techno-economical way. For proper design and operation of the wastewater treatment plant, high COD wastewater streams need to be identified and segregated.

COD in wastewater analysis is by online or lab method. COD can be measured in real-time with our COD analysers to improve wastewater process control and plant efficiency. For online COD monitoring, a total organic carbon (TOC) analyzer can be used with or without transmitter facility. The relationship between TOC analyzer and COD for respective COD value was set in the analyzer.

To make online measurement, COD to TOC ratio should be established; this varies from two to five based on effluent nature. COD analysis by lab method is described below.

2.2.6.1.1 Reagents

- Potassium dichromate 0.25 N: Dissolve 12.258 g potassium dichromate in distilled water and make up to 1000 ml in a volumetric flask.
- Ferroin indicator.
- Ferrous ammonium sulfate (FAS): Dissolve 98 g FAS in a little water, add 20 ml conc. H_2SO_4, and make it up to 1000 ml.
- Silver sulfate.
- Mercuric sulfate.

2.2.6.1.2 Standardization of FAS

- Pipette 10 ml 0.25 N potassium dichromate in a volumetric flask, make it up to 100 ml, and transfer to a conical flask. Add 30 ml conc. sulfuric acid and cool to room temperature. Add three drops of ferroin indicator and titrate with FAS.

N1V1 = N2V2

N1 = Normality of potassium dichromate, i.e. 0.25 N.

V1 = Volume of potassium dichromate, i.e. 10 ml

N2 = Normality of FAS

V2 = Volume of FAS required for titration

2.2.6.1.3 Procedure

Shake the sample well. Make dilutions: for inlet, 1 : 250, i.e. 1 ml sample in 250 ml volumetric flask, and make up to the mark; for outlet, treated 1 : 100 dilution, i.e. 1 ml sample in 100-ml volumetric flask, and make up to the mark. Place 10 ml potassium dichromate in flask. Add 20 ml diluted sample and 30 ml (approx.) conc. H_2SO_4. One pinch of mercuric sulfate and one pinch of silver sulfate is added and kept for refluxing for two hours.

After two hours remove the samples and cool to room temperature, add 80 ml distilled water and three to four drops of ferroin indicator, and titrate with standard FAS. End point is green to wine-red color.

ml of FAS required for titration = 'A'.

Conduct a blank using distilled water in place of sample (the quantities of other reagents added are the same as that added for the sample).

ml of FAS required for titration = 'B' for blank.

2.2.6.1.4 Calculation

$$COD(mg/l) = \left[\frac{(Blank\ titre\ value - sample\ titre\ value) \times}{(Normality\ of\ FAS \times 8000 \times dilution)} \right] / ml\ of\ sample\ taken$$

2.2.7 Biological Oxygen Demand

Biological oxygen demand (BOD) value is a measurement of the amount of DO that is used by aerobic microorganisms when decomposing organic matter in wastewater. For design of wastewater treatment plant, BOD to COD ratio should be evaluated for proper selection of treatment units. If ratio is >0.4, wastewater is easily degradable through biological treatment; if ratio is <0.2, it is hard to degraded.

BOD in wastewater analysis is by online or lab method. BOD can be measured in real-time with a BOD analyser to improve wastewater process control and plant efficiency. For online BOD monitoring a TOC analyzer can be used. The relationship between TOC analyzer and BOD for respective BOD value is set in the analyzer. To make an online measurement, BOD analysis should be established based on requirement only in view of the long analysis process. For daily operation, TOC and COD values are usually monitored. BOD analysis by lab method is described below.

2.2.7.1.1 Reagents

- Manganese sulfate solution: Dissolve manganese sulfate (480 g $MnSO_4 \cdot 4H_2O$, 400 g $MnSO_4 \cdot 2H_2O$, or 364 g $MnSO_4 \cdot H_2O$) in freshly boiled and cooled water, filter, and make up to 1000 ml. The solution should not give a blue color by addition of acidified potassium iodide solution and starch.

- Alkali-iodide azide reagent:
  - Dissolve 175 g potassium hydroxide (or 125 g sodium hydroxide) and 37.5 g potassium iodide (or 33.7 g sodium iodide) in distilled water and make up to 250 ml.
  - Dissolve 2.5 g sodium azide in 10 ml distilled water.
  - Pour the azide solution into the alkali-iodide solution and mix well.
- Concentrated sulfuric acid.
- Sodium thiosulfate 0.1 N: Dissolve 24.82 g sodium thiosulfate $Na_2S_2O_3 \cdot 5H_2O$ in boiled and cooled distilled water and make up to 1000 ml.
  - Sodium thiosulfate 0.025 N: Dilute 250 ml of the above solution to 1000 ml with distilled water.
- Starch solution indicator.
- To prepare an aqueous solution of starch, dissolve 2 g of laboratory grade soluble starch and 0.2 g salicylic acid, as a preservative, in 100 ml hot distilled water and cool.

2.2.7.1.2 Reagents for Dilution Water

- Calcium chloride solution: Dissolve 27.5 g anhydrous calcium chloride $CaCl_2$ in distilled water and dilute to 1000 ml.
- Magnesium sulfate solution: Dissolve 22.5 g magnesium sulfate heptahydrate $MgSO_4 \cdot 7H_2O$ in distilled water and dilute to 1000 ml.
- Ferric chloride solution: Dissolve 0.25 g ferric chloride hexa hydrate $FeCl_3 \cdot 6H_2O$ in distilled water and dilute to 1000 ml.
- Phosphates buffer solution: Dissolve 8.5 g potassium dihydrogen phosphate KH_2PO_4, 21.75 g dipotassium hydrogen phosphate, 33.4 g disodium hydrogen phosphate heptahydrate $Na_2HPO_4 \cdot 7H_2O$, and 1.7 g ammonium chloride in 500 ml distilled water and make up to 1000 ml. The pH of this buffer solution should be 7.2 and it should be keep it in the refrigerator.

2.2.7.1.3 BOD Dilutions

1 : 1000 3 ml of 1 : 10 in 300-ml BOD bottle
1 : 2000 1.5 ml of 1 : 10 in 300-ml BOD bottle
1 : 3000 1 ml of 1 : 10 in 300-ml BOD bottle
1 : 4000 0.75 ml of 1 : 10 in 300-ml BOD bottle
1 : 10000 3 ml of 1 : 100 in 300-ml BOD bottle

2.2.7.1.4 Preparation of Dilution Water

Take 1 ml of each of calcium chloride, magnesium sulfate, ferric chloride, and phosphate buffer solution and add 1000 ml water. Before addition of these chemicals aerate water for three to four hours and store at 27 °C in a BOD incubator with temperature always maintained at 27 °C. This is standard dilution water – prepare it first before use.

2.2.7.1.5 Sample Volume and Dilution Techniques

On the basis of COD, determine expected BOD. Use the following formula to calculate sample volume:

$$\frac{\text{Sample Volume in ml.,}}{\text{per liter of dilution}} = \frac{X}{\text{Expected BOD}} \times 1000$$

For keeping two dilutions take X = 2.5 and 4.0.

For single dilution take X = 3.0 or 3.5.

Round off to nearest convenient volume fraction.

In the case of high BOD samples, prepare primary dilutions with distilled water and then make the final dilution.

2.2.7.1.6 Procedure

- Collect the sample in a BOD bottle and fill it with dilution water up to the mouth.
- Add 2 ml manganese sulfate and 2 ml alkali-iodide azide solution. Stopper the bottle and mix it by inverting 10 times. The tip of the pipette should be below the surface of liquid.
- Allow the precipitate to settle completely, leaving a clear supernatant liquid.
- Carefully remove the stopper and add 2 ml conc. sulfuric acid by pouring down the sides of the bottle.
- Stopper the bottle and mix thoroughly until dissolution is complete.
- Measure 203 ml of the solution from the bottle into a conical flask of 500 ml capacity.
- Titrate it immediately with 0.025 N sodium thiosulfate solution, using starch as the indicator.
- For all the samples take two sets: one should be titrated immediately and the other kept in an incubator for three or five days at 27 °C and its DO titrated.
- Carry out a blank with each set using water instead of sample.

2.2.7.1.7 Calculation

$$BOD(mg/l) = \left[(Do - D5) - (Bo - B5) \right] \times Dilution / ml \text{ sample volume}$$

where,

D0 = Initial DO

D5 = After three or five days DO

B0 = Initial DO for blank

B5 = After three or five days DO for blank

Note: During addition of any reagent there should be no bubble formation in the BOD bottle.

2.2.8 Total Kjeldhal Nitrogen

Total Kjeldhal nitrogen (TKN) analysis gives the total value of ammoniacal nitrogen and organic nitrogen in wastewater. Organic nitrogen compounds in wastewater are converted into NH_3 and ammonium ion (NH^+_4) by microbial activity. Ammonium ion is the first inorganic nitrogen species produced during biological wastewater treatment. Nitrification is a two-step biological process used to remove ammonium in wastewater. The bacterium

Nitrosomonas converts ammonium to nitrite (NO_2). The bacterium *Nitrobacter* converts nitrite to nitrate (NO^-_3).

Nitrogen in wastewater analysis is by online or lab method. For online nitrogen monitoring in wastewater, an ammoniacal analyzer can be used with or without transmitter facility.

TKN analysis by lab method is described below.

2.2.8.1.1 Reagents

- Concentrated sulfuric acid.
- Copper sulfate solution 10%: Dissolve 10 g copper sulfate, $CuSO_4{\cdot}5H_2O$ in 100 ml distilled water.
- Phenolphathalein indicator.
- Sodium hydroxide 50%: Dissolve 100 g NaOH in 200 ml distilled water.
- Boric acid solution 2%: Dissolve 10 g boric acid H_3BO_3 in ammonia-free distilled water and dilute to 500 ml.
- Mixed indicator solution: Dissolve 200 mg methyl red indicator in 100 ml 95% ethyl alcohol. Dissolve 100 mg methylene blue in 50 ml 95% ethyl alcohol. Combine the two solutions. Prepare monthly.
- Standard sulfuric acid solution 0.02 N: For 1 N H_2SO_4: 14 ml conc. H_2SO_4 and make it to 500 ml with distilled water. For 0.02 N H_2SO_4: 20 ml conc. H_2SO_4 and make it to 1000 ml with distilled water.

2.2.8.1.2 Procedure

Place 100 ml or an appropriate volume of the sample in a Kjeldhal flask. Add 10 ml conc. H_2SO_4, and 1 ml copper sulfate solution. If the organic matter is hard to destroy add 20 ml conc. H_2SO_4 and 5 g potassium sulfate. Add a few glass beads and boil, until the solution becomes clear. Then digest for an additional 30 minutes and allow to cool.

2.2.8.1.3 Distillation

Transfer the contents of the flask carefully into a distillation flask and dilute to about 300 ml. Make the solution alkaline with sodium hydroxide using phenolphthalein indicator. Start the distillation after immersing the tip of the condensor in 50 ml boric acid solution in a conical flask. Collect about 200 ml of the distillate.

2.2.8.1.4 Titration

Add 0.5 ml mixed indicator solution to the distillate. Titrate against 0.02 N H_2SO_4. End point is the color change from pale green to lavender. Conduct a blank also starting from the digestion step to final titration.

2.2.8.1.5 Calculation

$$TKN\left(mg/l\right)=\left[\left(\frac{ml\ 0.02\,N\ H2SO4\ for\ sample-}{ml\ 0.02\,N\ H2SO4\ for\ blank}\right)\times\left(0.28\times1000\right)\right]/ml\ of\ sample\ volume$$

2.2.9 Total Phosphorus

Phosphorus is one of the major nutrients contributing to the increased eutrophication of lakes and natural water bodies. Normally secondary biological treatment can only remove very low quantities of phosphorus, so excess is removed mostly through chemical precipitation. Total phosphorus includes all orthophosphates and condensed phosphates, both dissolved and particulate, organic and inorganic. To release phosphorus from combination with organic matter, digest and oxidize. After digestion, determine liberated orthophosphates by colorimetery.

2.2.9.1 Sulfuric Acid–Nitric Acid Digestion
2.2.9.1.1 Apparatus
- Digestion rack: An electrically or gas-heated digestion rack with provision for withdrawal of fumes is recommended. Digestion racks typical of those used for Kjeldahl digestion are suitable.
- Micro Kjeldahl flasks.

2.2.9.1.2 Reagents

- Conc. sulfuric acid.
- Conc. nitric acid.
- Phenolphthalein indicator.
- Sodium hydroxide 1 N.

2.2.9.1.3 Procedure
Into a micro Kjeldahl flask, measure a sample containing the desired amount of phosphorus. Add 1 ml conc. H_2SO_4 and 5 ml conc. HNO_3. Digest to a volume of 1 ml and then continue until solution becomes colorless to remove HNO_3.

Cool and add approx. 20 ml distilled water, 0.05 ml phenolphthalein indicator, and as much 1 N NaOH solution as required to produce a faint pink tinge. Transfer neutralized solution, filtering if necessary to remove particulate material or turbidity, into a 100-ml volumetric flask. Add filter washings to flask and adjust sample volume to 100 ml with distilled water.

Determine phosphorus by vanadomolybdophosphoric acid colorimetric method.

2.2.9.2 Vanadomolybdophosphoric Acid Colorimetric Method
2.2.9.2.1 Apparatus
Colorimetric equipment, one of the following:

- Spectrophotometer: For use at 400–43 nm.
- Filter photometer: Provided with the blue or violet filter exhibiting maximum transmittance between 400 and 470 nm.

The wavelength at which color intensity is measured depends on sensitivity desired, because sensitivity varies 10 times with wavelength 400–43″ nm. Concentration ranges for different wavelengths are shown in Table 2.32.

Table 2.32 Phosphorus range and wavelength.

Phosphorus (P) range (mg/l)	Wavelength (nm)
1.0–5.0	400
2.0–10.0	420
4.0–18.0	470

2.2.9.3 Acid Washed Glassware

Use acid washed glassware for determining low concentration of phosphorus. Phosphate contamination is common because of its absorption on glass surfaces. Avoid using commercial detergents containing phosphates. Clean all glassware with hot dilute HCl and rinse well with distilled water. Preferably, reserve the glassware only for phosphate determination, and after use, wash and keep filled with water until needed. If this is done, acid treatment is required only occasionally.

2.2.9.3.1 *Reagents*

- HCl: $1 + 1$ H_2SO_4, $HClO_4$, or HNO_3 may be substituted for HCl. The acid concentration in the determination is not critical, but a final sample concentration of 0.5 N is recommended.
- Activated carbon.
- Vanadate molybdate reagent:
  - Solution A: Dissolve 25 g ammonium molybdate in 300 ml distilled water.
  - Solution B: Dissolve 1.25 g ammonium metavanadate by heating to boiling in 300 ml distilled water. Cool and add 330 ml conc. HCl. Cool solution B to room temperature, pour solution A into solution B, mix, and dilute to 1 l.
- Standard phosphate solution: Dissolve in distilled water 219.5 mg anhydrous KH_2PO_4 and dilute to 1000 ml; 1 ml $= 50.0\,\mu g$ PO_4.

2.2.9.3.2 *Procedure*

- Sample pH adjustment: If sample pH is greater than 10, add 0.05 ml (1 drop) phenolphthalein indicator to 50 ml sample and discharge the red color with $1 + 1$ HCl before diluting to 100 ml.
- Color removal from sample: Remove excessive color in sample by shaking about 50 ml with 200 mg activated carbon in an Erlenmeyer flask for five minutes and filter to remove carbon. Check each batch of carbon for phosphate because some batches produce high reagent blanks.
- Color development in sample: Place 35 ml or less of sample, containing 0.05–1.0 mg P, in a 50-ml volumetric flask, add 10 ml vanadate molybdate reagent, and dilute to the mark with distilled water. Prepare a blank in which 35 ml distilled water is substituted for the sample. After 10 minutes or more, measure absorbance of sample versus a blank at a wavelength of 400–43″ nm, depending on the sensitivity desired. The color is stable for days and its intensity is unaffected by variation in room temperature.

- Preparation of calibration curve: Prepare a calibration curve by using suitable volumes of standard phosphate solution and proceeding as in (c) above. When ferric ion is low enough not to interfere, plot a family of calibration curves of one series of standard solutions for various wavelengths. This permits a wide latitude of concentrations in one series of determinations. Analyze at least one standard with each set of samples.

2.2.9.3.3 Calculation

$$P/l\left(mg/l\right)=\left[mg\,P\left(in\,50\,ml\,final\,volume\right)\times 1000\right]/ml\,of\,sample\,volume$$

2.2.10   MLSS and MLVSS

Mixed liquor suspended solids (MLSS) consists mostly of microorganisms and non-biodegradable suspended matter. MLSS is an important part of the activated sludge process to ensure that there is a sufficient quantity of active biomass available to consume the applied quantity of organic pollutant. This is known as the food to microorganism (F/M) ratio. By maintaining F/M ratio at the appropriate level the biomass will consume high percentages of the food. F/M ratio to be maintained in the range of 0.15–0.3 based on incoming effluent nature. MLSS is a combination of sludge and water removed from the clarifier in the wastewater treatment process and reintroduced into an earlier phase of the treatment process. The mixed liquor contains microorganisms which digest the wastes in the wastewater.

MLSS is a test for the TSS in a sample of mixed liquor. This test is essentially the same as the test for TSS in wastewater, except for the use of mixed liquor as the wastewater sample. MLSS should be maintained at more than 3000 ppm as per treatment technology requirement.

2.2.10.1.1   Procedure

- Collect a grab sample of mixed liquor.
- Measure the TSS. Record the sample volume and the combined sample and filter weight. At least 10% of all samples should be analyzed in duplicate.
- Ignite the filter and the TSS residue from step 1 in a muffle furnace at 550 °C. An ignition time of 15–20 minutes is usually sufficient for 200 mg residue.
- Let the filter cool partially in the air until most of the heat has dissipated. Then transfer the filter to a desiccator to cool the rest of the way to air temperature.
- Weigh the filter and record the weight.
- Repeat the cycle of igniting, cooling, desiccating, and weighing until a constant weight is obtained or until the weight change is less than 4% or 0.5 mg, whichever is less.

2.2.10.1.2   Calculation

$$mg\,total\,mixed\,liquor\,suspended\,solids/L =\left(A-B\right)\times 1000/ml\,of\,sample\,volume$$

where

A = weight of filter + dried residue (mg)

B = weight of filter (mg)

Mixed liquor volatile suspended solids (MLVSS) is determined by taking the dried MLSS sample and firing it in an oven at 550 °C. The ash is cooled in a desiccator and then weighed. The mg/l of non-volatile ash can be calculated the same way as MLSS. The MLVSS is calculated by subtracting the non-volatile ash number from the MLSS.

Note: The total weight of MLSS within an aeration tank can be calculated by multiplying the concentration of MLSS (kg/m^3) in the aeration tank by the tank volume (m^3).

2.2.11 Sludge Volume Index

Sludge volume index (SVI) is used to monitor settling characteristics of activated sludge and other biological suspensions in the aeration tank in the activated sludge process. It is a process control parameter to determine the recirculation rate of sludge. SVI is the volume in milliliters occupied by 1 g of a suspension after 30 minutes settling. Although SVI is not supported theoretically, experience has shown it to be useful in routine process control.

2.2.11.1.1 Procedure

- Determine the suspended solids concentration of a well-mixed sample of the suspension.
- Determine the 30 minutes settled sludge volume.
- Measure the amount of settleable solids from the scale of the measuring cylinder. This settleable material is known as "settleable sludge volume index" (SSVI).

2.2.11.1.2 Calculation

$$SVI(mg/l) = SSVI(ml/l) \ in \ 30 \ min. * 1000 / MLSS(mg/l)$$

2.2.12 Turbidity

Turbidity in wastewater is caused by suspended matter, such as clay, silt, finely divided organic and inorganic matter, soluble colored organic compounds, and plankton and other microscopic organisms. Turbid water has a muddy or cloudy appearance and is esthetically unattractive. To remove turbidity from wastewater, coagulation and the flocculation process are used. The pretreatment section of the wastewater treatment plant should be properly design based on inlet turbidity value and considering other parameters such as TSS and O&G.

Turbidity in wastewater analysis is by online or lab method. For online turbidity monitoring a turbidity analyzer can be used with or without transmitter facility. Turbidity analysis by lab method is described below.

2.2.12.1.1 Outline of Method
The sample is matched against standard suspensions of Fuller's Earth in water.

2.2.12.1.2 Terminology
For the purpose of this test, the following definition shall apply: Scale Unit – Turbidity imparted by 1 mg of Fuller's Earth when suspended in 1000 ml of distilled water.

2.2.12.1.3 Preparation of Turbidity Standards
Mix slowly with constant stirring 5.0 g of Fuller's Earth previously dried and sieved through a 75-μm IS (International Standard) sieve with distilled water and dilute it to 1000 ml. Agitate

intermittently for one hour and then allow to stand for 24 hours. Withdraw the supernatant liquid without disturbing the sediment. Evaporate about 50 ml of the removed liquid, dry the residue at $105 + 2\,°C$, and weigh the residue to determine the amount of clay in suspension. Prepare turbidity standards with this standardized stock suspension with distilled water. A drop of saturated mercuric chloride solution may be added as preservative. The standards are stable for three months.

2.2.12.1.4 Procedure

After thorough shaking, pour the sample into a clear glass bottle of suitable capacity, 1 l. Compare it with the turbidity standards contained in similar bottles, holding them against a suitable background and using a source of light, which illuminates them equally and is placed so that no rays reach the eye directly. The sample and the standards should be shaken simultaneously immediately before comparison. If the sample has turbidity over 100 units, dilute it with distilled water before testing and multiply the result with an appropriate factor.

2.2.12.1.5 Note

Comparison of turbidity may also be done with the help of suitable instruments.

2.2.13 Color

Wastewater can have impurities of various colored compounds (blue, green, redish, whitish, blackish, yellowish, etc.) Color analysis by lab method is described below.

2.2.13.1.1 Outline of Method

The color of the sample is matched against a series of standards containing potassium chloroplatinate and cobalt chloride.

2.2.13.1.2 Terminology

For the purpose of this test, the following definitions shall apply:

- True color: Color due to substances in solution, after removal of suspended matter.
- Apparent color: Color due to substances that are in solution as well as in suspension.
- Hazen unit: Color obtained in a mixture containing either 1 mg platinum or 2.49 mg potassium chloroplatinate along with 2 mg cobalt chloride ($CoCl_2 \cdot 6H_2O$) in 1 l of the solution.

2.2.13.1.3 Apparatus

A flat-bottomed Nessler tube of thin colorless glass and two types of tubes. The longer tubes should be 45 cm tall and 2.5 cm internal diameter. The shorter tubes should be 30 cm tall and 1.7 cm internal diameter. Tubes of any one type should be identical in shape, and the depth measured internally from the graduation mark to the bottom should not vary by more than 2 mm in the tubes used.

2.2.13.1.4 Reagents

- Platinum or potassium chloroplatinate aqua regia – prepared by mixing one part by volume of concentrated nitric acid (conforming to IS 264–1950) with three parts by volume of concentrated HCl (conforming to IS 265–1962).
- Crystalline cobalt chloride with the molecular composition ($CoCl_2 \cdot 6H_2$).

2.2.13.1.5 Procedure
2.2.13.1.5.1 Preparation of Color Standards
Dissolve 0.5 g metallic platinum in aqua regia and remove nitric acid by repeated evaporation to dryness in a water-bath after addition of excess of concentrated HCl (conforming to IS 265–1962). Dissolve the residue with 1.0 g of cobalt chloride in 100 ml of concentrated HCl to obtain a bright solution, if necessary by warming. Dilute the solution to 1000 ml with distilled water. This stock solution has a color of 500 Hazen units. A more convenient way of preparing the same solution is by dissolving 1.245 g of potassium chloroplatinate and 1.0 g of cobalt chloride in distilled water and diluting to 1 l. Prepare a set of color standards having color 5, 10, 15, 20, 25, 30, 35, 40, 50, 60, and 70 Hazen units by diluting the stock solution with water. Protect these color standards from evaporation and contamination when not in use. The color standards should be freshly prepared for each determination. But in routine practice, they may be used repeatedly, provided they are protected against evaporation and contamination when not in use.

2.2.13.1.5.2 Procedure for Clear Samples
For samples having turbidity under 5 mg/l, match the color of the sample against the standard colors in the longer Nessler tubes. Fill the tubes to mark and compare the color by looking vertically downwards against a pure white surface. If the color is found to exceed 70 units, dilute the sample with distilled water before comparison and multiply the result by the appropriate factor. As matching is very difficult when the color of the sample is below 5 Hazen units, report the color as less than 5 Hazen units in such cases. When the color of the sample exceeds 30 Hazen units, the comparison may, if desired, be made in the shorter Nester tubes.

2.2.13.1.5.3 Procedure for Turbid Samples
If the sample has turbidity over 5 mg/l it becomes impossible to measure the true color accurately by the method prescribed in the procedure for clear samples, and if an attempt is made, the value found shall be reported as "apparent color." In the presence of turbidity, the true color shall be determined after centrifuging. The sample should be centrifuged until the supernatant liquid is clear. The centrifuged clear sample should be compared by the method prescribed in the procedure for clear samples.

2.2.13.1.6 Note
For estimating true color, filter paper should not be used since this leads to erroneous results. The color determination should be made as early as possible after the collection of samples as certain biological changes occurring in storage may affect the color.

2.2.13.1.7 Report
The results of color determination should be excess in whole numbers and recorded as follows:

- 1–50 reports to the nearest 1 Hazen unit.
- 51–100 reports to the nearest 5 Hazen units.
- 101–250 reports to the nearest 10 Hazen units.
- 251–500 reports to the nearest 20 Hazen units.

2.2.14   Temperature

Adaptability in microorganisms is stable up to a certain temperature range. After the range of temperature 35–40 °C, bacterial performance and adaptability is affected, and it may also lead to a diminished bacterial colony. So temperature needs to be monitored in the wastewater treatment plant.

Temperature measurements may be made with any good mercury-filled Celsius thermometer. As a minimum, the thermometer should have a scale marked for every 0.1 °C, with markings etched on the capillary glass. The thermometer should have a minimum thermal capacity to permit rapid equilibration. Periodically check the thermometer against a precision thermometer certified by the National Institute of Standards and Technology (NIST), formerly the National Bureau of Standards, which is used with its certificate and correction chart. For field operations use a thermometer having a metal case to prevent breakage. Dip the calibrated thermometer just above the solution surface in the measuring glassware/volumetric flask and take readings on the scale carefully to determine the temperature.

2.2.15   Acidity

2.2.15.1.1   Reagents
0.12909 N NaOH: Dissolve 4 g NaOH in water and make it up to 1000 ml.

2.2.15.1.2   Procedure
Take a 100-ml sample in a beaker, note its initial pH, and slowly add 0.1 N NaOH to bring its pH to 8.3. Note the ml of 0.1 N NaOH required to bring its pH to 8.3.

2.2.15.1.3   Calculation

$$Acidity\,(mg\,/\,l) = \left(ml\;0.1\;N\;NaOH \times Normality \times 50000\right)/\,ml\;of\;sample\;volume$$

2.2.15.1.4   Note
Acidity increases with the increase in volatile acid content.

2.2.16   Alkalinity

2.2.16.1.1   Reagents
1 N H_2SO_4: 14 ml conc. sulfuric acid made up to 500 ml in a volumetric flask.

2.2.16.1.2   Procedure
Take a 100-ml sample in a beaker and note its initial pH with pH meter. Add 1 N H_2SO_4 to bring its pH to 3.7 by adding 1 N H_2SO_4 slowly. Note the ml of 1 N H_2SO_4 used in bringing the pH to 3.7.

2.2.16.1.3   Calculation

$$Total\;Alkalinity\,(mg\,/\,l) = \left(ml\;of\;H2SO4 \times N\;of\;H2SO4 \times 50000\,ml\right)/\,ml\;of\;sample\;volume$$

2.2.17 Hardness

Hardness in water is due to the presence of bicarbonates, chlorides, and sulfates of calcium and magnesium. Temporary hardness is due to the presence of bicarbonates and permanent hardness is due to the presence of chlorides and sulfates. Sometimes, hardness may include iron, aluminum, zinc, manganese, etc.

Hardness in wastewater analysis is by online or lab method. For online hardness monitoring a hardness analyzer can be used with or without transmitter facility. The lab method is described below.

2.2.17.1 Method A: Complexametric Method (EDTA Method)

Ethylene diamine tetra acetic acid (EDTA) forms a chelated soluble complex when added to a solution of certain metal ions. If a small amount of Eriochrome black T (EBT) is added to an aqueous solution containing calcium and magnesium ions at pH of $10.0 + 0.1$, the solution will become red-wine in color. If EDTA is then added as a titrant, the calcium and magnesium will be complexed. After sufficient EDTA has been added to complex all the calcium and magnesium, the solution will turn blue from red-wine.

2.2.17.1.1 Reagents

- Ammonia buffer of pH 10.
- Solution of disodium salt of EDTA.
- EBT indicator solution.

2.2.17.1.2 Procedure

Take a 50-ml of sample in an Erlenmeyer flask, add four to six drops of EBT indicator solution, add 1 ml of buffer solution, and mix. Titrate immediately with EDTA solution till the color changes from red to blue.

2.2.17.1.3 Calculation

$$\text{Total hardness as } CaCO3 \ (mg/l) = \text{Volume of } 0.01M \ EDTA \ soln. \times 1000/ml \ of \ sample \ volume$$

2.2.17.1.4 Note

For checking hardness in soft water use 500 ml of the sample in a 750-ml evaporating dish and add 3 ml of buffer solution followed by 10–12 drops of indicator solution.

2.2.17.2 Method B: Soap Solution Method

This is a quick method for checking hardness of treated water or an accurate determination of the hardness of treated water; the EDTA method for hardness should be used.

2.2.17.2.1 Reagents

- "B" soap solution.
- 40-ml shaking bottle.

2.2.17.2.2 Procedure

Take a water sample up to the 40-ml mark in a shaking bottle. Add 10 drops of "B" soap solution. Shake vigorously. If lather is obtained which will last for one to two minutes, the water is soft. If no lather is obtained, or if the lather does not last, the water is hard.

2.2.17.2.3 Note

Thoroughly rinse the shaking bottle clean with soft water. The soap solution bottle should be kept tightly stopper; otherwise the solution will evaporate and give a false reading.

2.2.17.3 Calcium Hardness

The water sample is titrated against EDTA solution using murexide indicator (ammonium purpurate) in highly alkaline medium.

2.2.17.3.1 Reagents

- Approximately 1 N sodium hydroxide solution.
- Standard EDTA solution.
- Murexide indicator.

2.2.17.3.2 Procedure

Prepare a color comparison blank in a white porcelain basin by stirring 2.0 ml of 1 N NaOH, 0.2 g solid indicator mixture (or four to six drops of indicator solution) into 50 ml of distilled water and sufficient EDTA titrant (0.05–0.1 ml) to produce an unchaining purple color. Pipette into a similar basin 50 ml of sample, neutralize the alkalinity with 0.02 N HCl, boil for two to three minutes to expel the CO_2, and cool to room temperature. Add 2.0 ml 1 N NaOH, or a volume sufficient to produce a pH of 12–13, and mix. Add 0.2 g of powdered indicator. Stirring constantly, titrate with standard EDTA solution to the color of the comparison blank. Check the end point by adding one or two drops of titrant in excess to be sure that no further deepening of the purple color takes place.

2.2.17.3.3 Calculation

$$Calcium\ as\ CaCO3 (mg/l) = \left[(A - B) \times C \times 1000 \right] /ml\ of\ sample\ volume$$

where

A = EDTA required for titration of sample in ml
B = EDTA required for blank in ml
C = $CaCO_3$ equivalent to 1.0 ml of EDTA in mg

(See water hardness scale in Table 2.33.)

Table 2.33 Water hardness scale.

Concentration (ppm)	Hardness rating
<61	Soft
61–120	Moderately hard
121–180	Hard
>180	Very hard

2.2.17.3.4 Note

The only serious interference with EDTA titrant of calcium is that of orthophosphate ion. If calcium hardness exceeds about 60 ppm $CaCO_3$ and concentration of orthophosphate is 10 ppm or more, calcium phosphate is precipitated when pH is raised to 12, giving low results. Phosphate, if present, can be removed by ion exchange.

2.2.18 Iron

Iron is frequently found in natural waters. In addition to the "natural" iron content of water, iron is passed into solution when the corrosion of iron and steel surface occur. Iron analysis by lab method is described below.

2.2.18.1 Method A: O-Phenanthroline Method

Calorimetric estimation of iron is based on formation of the orange-red phenanthroline complex (C12 H2 N2) 3 Fe^{2+} in the pH range 2–9. Below pH 2 the color develops slowly and is much weaker. Since this complex is formed with ferrous iron hydroxylamine, hydrochloride is the most satisfactory reducing agent (for the reduction of ferric iron). Certain divalent metals such as Cd, Hg, and Zn form slightly soluble complexes with the reagent and reduce the intensity of the iron color, but this interference may be minimized by adding a large excess of reagent. Phosphorus may be present up to 20 ppm. Fluoride up to 500 ppm does not interfere if the pH is kept above 4.0.

2.2.18.1.1 Reagents

- Standard iron solution.
- 1–10 phenanthroline monohydrate reagents.
- Ammonium acetate buffer solution.
- Hydroxylamine hydrochloride.

2.2.18.1.2 Procedure

For total iron use a well-mixed sample. For determination of dissolved iron if precipitated iron is present, decant a sample, allow settling the precipitations to settle down, and filter. If precipitated iron is present, use the filtrate. Prepare a series of visual standards/photo-metric calibration curve by measuring the following amounts of standard iron solutions into beakers: 0.01, 0.02, 0.03, 0.04, 0.05, up to 0.1 mg. Dilute the solution to 50 ml. To the sample, blank, and the standards add 2 ml conc. HCl, 1 ml hydroxylamine solution, and glass beads, and boil until the volume is reduced to 15–20 ml. Transfer the solution to Nessler tubes if visual comparison is to be made or to 50- or 100-ml volumetric flasks if the spectrophotometer method is used. Add 10 ml ammonium acetate buffer and 5 ml 1–10 phenanthroline reagent, dilute to the mark with distilled water, and mix well. After 10–15 minutes compare the color visually or measure the absorbance at 510 mm. The color forms in the range of pH 2–9 and is very stable.

2.2.18.1.3 Calculation

$$Iron \ as \ Fe \left(mg \ / \ l \right) = \left(mg \ of \ Fe \times 1000 \right) / \ ml \ of \ sample \ volume$$

2.2.18.2 Method B: Iron Analysis by Thioglycolic Acid Method

In ammoniacal medium, mercaptoacetate (as the ammonium salt $HSCH_2 COONH_4$) reacts with iron to yield the soluble red-purple product $Fe (OH) (SCH_2COO)_2$ containing iron in the ferric state. Ferrous iron reacts to give the same complex by air oxidation. In the absence of oxygen, the color slowly fades as a result of the reduction of ferric iron to ferrous and oxidation of mercaptoacetate to the disulfide – $O_2 CCH_2 SSCH_2CO_2$. The air oxidation of ferrous iron in the system is very rapid, and usually under analytical conditions, more than the pace of the reduction so that a stable coloration is obtained. The reaction must be carried out in basic medium; adding citrate may prevent precipitation of metal hydroxides. Certain metals, among them Cu (21 mg), arsenic +3 (>100 mg), tin +2, zinc (>10 mg), and cadmium black, affect the iron color, although this effect can be diminished by the addition of more reagent. Very high concentration of salts of alkali metals decreases the color intensity somewhat. On the other hand, anions have but a slight effect on the color. As much as 5000 ppm chloride, fluoride, orthophosphate, tartrate, oxalate, and citrate do not interfere. The color is fairly stable (at least six hours in diffuse light).

2.2.18.2.1 Reagents

- Thioglycolic acid (80%).
- Ammonia solution (sp. Gr. 0.91).

2.2.18.2.2 Procedure

Place 100 ml of water sample in a 250-ml beaker. Add 2.5 ml thioglycollic acid reagent. After stirring, evaporate in sand or water-bath to 5 ml (not less) and then allow cooling. Add 5.0 ml ammonia solution, shake, and pour into a 25-ml measuring flask. Make up to the mark. Measure optical density at 540 mm using a 5-cm cell. Standard with reagent blank is carried simultaneously.

2.2.18.2.3 Calculation

$$Iron\ as\ Fe\left(mg\,/\,l\right)=\left(mg\ Fe\times 1000\right)/\ ml\ of\ sample$$

2.2.19 Silica

The silica content of natural water will vary to a considerable extent depending on the locality. The presence of silica is particularly objectionable in water used for boiler feed purpose as it may lead to the formation of hard, dense scale. In addition, a very serious problem encountered in high-pressure operations is the deposition of siliceous materials on turbine blades and super heaters.

Silica content in wastewater analysis is by online or lab method. For online silica content monitoring a silica analyzer can be used with or without transmitter facility. The lab method is described below.

The gravimetric method is the standard applicable above 20 mg/l SiO_2 content. This method should be followed for standardization of standard silicate solution used in colorimetric methods. The heterophony blue colorimetric method is adaptable for the range of 0–2 mg/l SiO_2 and the yellow molybdate silicate method in the range of 0–20 mg/l. Reagent blank should always be used in all three methods.

2.2.19.1 Method A: Gravimetric Method

2.2.19.1.1 Procedure

Take a sample containing at least 10 mg SiO_2. If necessary clarify by filtration. Acidify with 2 or 3 ml conc. HCl and evaporate to dryness in a platinum dish in a water-bath. At regular intervals add two or more portions of 2–3 ml conc. HCl as additional quantities of sample are added to the dish. Bake the evaporated residue in an oven at 110 °C for about an hour. Add 5 ml conc. HCl, warm, and add 50 ml distilled water. Loosen the clinging residue from the sides and bottom of the dish and filter, collecting the filtrate. Wash the dish and residue with hot 1 : 50 HCl and finally with distilled water until the washings are free from chloride. Return the filtrates and washings to the platinum dish and again evaporate to dryness. Repeat as previously, collecting the residue in another filter paper. Dry the two filter papers with residue, burn, ignite at 1000–1200 °C in a platinum crucible, and weigh. Moisten the residue with a few drops of distilled water; add two drops of H_2SO_4 and 10 ml 48% hydrogen fluoride (HF). Cautiously evaporate to dryness in a steam-bath in a fume cupboard. Again ignite at 1000–1200 °C, cool, and weigh. Carry out a blank.

2.2.19.1.2 Calculation

$$Silicon\ dioxide\ as\ SiO2\ (mg\ /\ l) = (A - B) - (C - D) \times 1000\ /\ ml\ of\ sample$$

where

A = Weight of crucible and sample residue in mg after first ignition

B = Weight of crucible and sample residue in mg after HF treatment and second ignition

C = Weight of crucible and blank residue in mg after first ignition

D = Weight of crucible and blank residue in mg after HF treatment and second ignition

2.2.19.2 Method B: Colorimetric Estimation of Reactive Silica

Ammonium molybdate at approximately pH 1.2 reacts with silica and any phosphate present to produce heterophony acids. Oxalic acid is added to destroy the molybdophosphoric acid but not the molybdosilicic acid. Even if phosphate is known to be absent, the addition of oxalic acid is highly desirable and is a mandatory step. The intensity of the yellow color is proportional to the concentration of molybdate- reactive silica. The yellow molybdosilicic acid is reduced by means of 1 amino-2-naphthol-4-sulfonic acid to heterophony-blue. The blue color is more intense than the yellow color and provides increased sensitivity. In at least one of its forms, silica does not react with molybdate even though it is capable of passing through filter paper and is not noticeably turbid. The presence of such a molybdate unreactive silica is undesirable in raw water. It will not be removed in the water treatment plant and will find its way to the high-pressure stream system, where it will be converted to " molybdate-reactive" silica. Such increase in silica content will give rise to scale problems.

Chromate and large amounts of Fe, PO_4, sulfide, tannin, color, and turbidity are potential interferences. Oxalic acid treatment suppresses PO_4 and reduces tannin interference. Inorganic sulfide can be removed by boiling an acidified sample. The addition of 1 ml of 1% EDTA solution after molybdate reagent overcomes high Fe and Ca concentrations.

2.2.19.3   Colorimetric Estimation of Silica 0–20 ppm SiO_2
2.2.19.3.1   Reagents
- Ammonium molybdate solution.
- N sulfuric acid.
- 10% oxalic acid.
- Lovibond comparator with standard silica disc.

2.2.19.3.2   Procedure
Fill one of the Nessler tubes to the 50-ml mark with sample and place in the left-hand compartment of the Lovibond comparator. Fill the other Nessler tube with 50 ml of sample at 25–30 °C. Add 2 ml of acidified ammonium molybdate solution. Mix thoroughly, add 4 ml of oxalic acid, and again mix thoroughly. Place in the right-hand compartment and allow to stand for 10 minutes. Stand the comparator facing a uniform source of light; compare the color of the sample in the disc. Rotate the disc until the colors are matched.

2.2.19.3.3   Calculation

$$Silicon\ dioxide\ as\ SiO2\ (mg\ /\ l) = Disc\ reading \times 20$$

2.2.19.3.4   Note
Should the color in the test solution be deeper than the deepest standard, a fresh test should be carried out using a smaller quantity of sample and diluting to 50 ml with distilled water before adding the reagents.

2.2.19.3.5   Silica-Free Water
Distilled water from an all-metal "still" or water that has been passed successively through a mixed bed deionization unit and strongly basic anion exchanger such as strong base anion resin regenerated with a regeneration level of 320 g/l NaOH has been found to be suitable. Prepare and store in a polyethylene bottle a large batch of water containing not more than 0.005 ppm SiO_2. Determine the silica content of water by treating it as a sample. This water is used to prepare reagents and standards and to dilute samples when necessary.

2.2.19.4   Colorimetric Estimation of Silica 0–2 ppm SiO_2
2.2.19.4.1   Reagents
- Acidified ammonium molybdate solution.
- 10% oxalic acid.
- Amino-naphthol reducing agent.
- Lovibond comparator with standard silica disc or spectrophotometer suitable for measurement at 815 microSiemens wavelength.

2.2.19.4.2   Procedure
Fill one of the Nessler tubes to the 50-ml mark with sample and place in the left-hand compartment of the Lovibond comparator. Fill the other Nessler tube with 50 ml of sample at 25–30 °C. Add 2 ml of acidified ammonium molybdate solution. Mix thoroughly and stand for five minutes. Add 4 ml of oxalic acid and mix well, then 2 ml of reducing agent, and mix well. Compare with that of a blank comprising the same water without reagents, using a

LoBiondo comparator or read the absorbance using a spectrophotometer (wavelength 815 microSiemens). Compute the silica content from the standard graph prepared from the standard silica solution.

2.2.20 Sulfate

2.2.20.1 Method A: Gravimetric Method

Sulfate is precipitated as barium sulfate by the addition of barium chloride to a slightly acidified solution of the sample at the boiling temperature. The precipitate is ignited and the residue is weighed as barium sulfate. The presence of silica and other non-volatile suspended matter causes high results. Sulfites and iron should also be present.

2.2.20.1.1 Reagents

- HCl 1 : 1.
- Methyl red indicator solution.
- Barium chloride solution.

2.2.20.1.2 Procedure

Take a sample containing approx. 50 mg sulfate. Dilute to 250 ml. Adjust the acidity with HCl to pH 4.5–5.0 using methyl red indicator. Add an additional 1–2 ml of HCl. Heat the solution to boiling and stir gently, add warm $BaCl_2$ solution slowly until precipitation appears to be complete, then add about 2 ml in excess. Digest the precipitate in a water-bath at 80–90 °C for not less than two hours. Filter through filter paper or through a Gooch crucible and wash with hot distilled water until washings are free from chlorides. Dry the filter paper and precipitate and ignite at 800 °C for one hour; do not let the filter paper flame. Cool in a desiccator and weigh.

2.2.20.1.3 Calculation

$$Sulphate\ as\ SO4\left(mg\ /\ l\right) = mg\ BaSO4 * 411.5\ /\ ml.\ of\ sample\ volume$$

2.2.20.2 Method B: EDTA Method

A measured excess of standard barium chloride solution is added to the sample and the excess barium chloride estimated by titration against standard EDTA solution.

2.2.20.2.1 Reagents

- Approx. 1 N nitric acid.
- Barium chloride of standard solution.
- pH 10 buffer solution.
- EBT indicator.
- EDTA solutions.

2.2.20.2.2 Procedure

Neutralize 100 ml of the sample with dilute nitric acid, adding a slight excess, and boil off to expel carbon dioxide. Add 10 ml or more if required of standard barium chloride solution to the boiling solution and allow it to cool. Dilute to 200 ml, mix, and allow precipitate to settle. Withdraw 50 ml of the supernatant liquid; add 0.5–1.0 ml of buffer and several drops of indicator solution. Titrate with standard EDTA solution to a blue color, which does not change on addition of further drops of EDTA solution.

2.2.20.2.3 Calculation

$$\text{Sulphates as } SO4 \, (mg / l) = 9.6 \left(0.1A + B - 4C \right)$$

where

A = Total hardness of sample (as $CaCO_3$ in mg/l)
B = Volume of standard barium chloride solution added in ml
C = Volume of standard EDTA solution required for titration in ml

2.2.20.2.4 Note

It is very difficult to judge the end point of titration of barium against EDTA using EBT as indicator. It is preferable to use standard $MgCl_2$ solution along with BaCl2 for $BaCl_2$ standardization (or use a mixture of $BaCl_2 + MgCl_2$ solutions, instead of $BaCl_2$ solution for precipitate of sulfate ion). It is also desirable to add $MgCl_2$ solution whenever sample is low in Mg ions, as in the case of decationized water.

2.2.21 Chloride

2.2.21.1 Method A: Silver Nitrate Method

Chloride is determined by titration with standard silver nitrate solution in the presence of potassium chromate indicator at neutral pH. Silver chloride is precipitated, and at the end point red silver chromate is formed.

2.2.21.1.1 Reagents

- N standard silver nitrate solution.
- Potassium chromate indicator.
- Phenolthalein indicator solution.
- N nitric acid solution.
- Calcium carbonate.

2.2.21.1.2 Procedure

Take 50 ml or 100 ml of sample in an Erlenmeyer flask. Add five drops of phenolthalein. If the sample turns pink, neutralize with 0.02 N nitric acid. If acidic (as in the case of decationized water) add a small amount of calcium carbonate. Add 1 ml of potassium chromate indicator and titrate with standard silver nitrate solution with constant stirring until there is a perceptible reddish color.

2.2.21.1.3 Calculation

$$\text{Chloride as } CaCO3 (m / l) = ml \; AgNO3 \times 1000 / ml \; of \; sample$$

2.2.21.1.4 Note

If sample is highly colored, add $(OH)_3$ suspension, mix, then let it settle. Filter and combine filtrate and washing. If sulfide, sulfite, and thiosulfate are present, add 1 ml H_2O_2 and stir for one minute. Bromide, iodide, and cyanide register as equivalent chloride concentration.

2.2.21.2   Method B: Titrated Method

Chloride ion can be titrated with mercuric nitrate because of the formation of soluble, slightly dissociated mercuric chloride. In the pH range 2.3–2.8 diphenylcarbozone indicates the end point of this titration by formation of purple complex with excess of mercuric ions. To keep the solution in the pH range, ± 0.1 pH unit nickel nitrate with HNO_3 is added.

2.2.21.2.1   Reagents

- 0.02 N mercuric nitrate standard solution.
- Diphenyl-carbozone indicator solution.
- Nickel nitrate solution.

2.2.21.2.2   Procedure

Take 50 ml of sample and add 1 ml of diphenylcarbozone indicator solution. Add 1 ml of nickel nitrate solution. Titrate against standard mercuric nitrate solution to color change of green to violet. Carry out blank titration with demineralized water.

2.2.21.2.3   Calculation

$$Chloride\ as\ CaCO3\ (mg\ /\ l) = (A - B) * 1000\ /\ ml.\ of\ sample\ volume$$

where

      A = Titrant required for sample in ml
      B = Titrant required for blank in ml

2.2.21.3   Method C: Chlorometric Method for Chloride (Mercury Thiocynate Method)

This method depends on the displacement of thiocynate ion from mercuric thiocynate by chloride ion. In the presence of ferric ion, a highly colored ferric thiocynate complex is formed and the intensity of its color is proportional to the original chloride ion concentration.

2.2.21.3.1   Reagents

- Mercuric thiocynate solution.
- Ferric ammonium sulfate solution.
- Standard sodium chloride solution: 0.1 mg Cl/ml.

2.2.21.3.2   Procedure

Take 10–20 ml of the water sample containing about 40 µg of chloride in a 25-ml standard flask. Add 2.0 ml of ferric ammonium sulfate solution, followed by 2.0 ml of mercuric thiocynate solution. Make up to the volume. Measure the optical density at 460 microSiemens from the calibration curve for the standard and compute the value for the sample.

2.2.21.3.3   Calculation

$$Chloride,\ as\ Cl\ (mg\ /\ l) = \mu g\ of\ chloride\ from\ graph\ /\ ml\ of\ sample$$

2.2.22   Residual Chlorine

2.2.22.1   Orthotolidine–Arsenite Method

Chlorine rapidly oxidizes orthotolidine (3,3-dimethylbenzidine) to the corresponding hol-oguinone, which is intensely yellow. This provides a sensitive test for available chlorine. The test is subject to interference by production of false color for chlorine when nitrite, iron, or manganese is present in the water. The false color is produced with orthotolidine by interfering substance present in the presence of sodium arsenite but the color does not appear due to residual chlorine. The test permits the measurement of the relative amounts of free available chlorine, combined available chlorine, and color due to interfering substances. The temperature of the sample should never be above 20 °C.

2.2.22.1.1   *Reagents*

- Phosphate buffer solution 0.5%.
- Potassium dichromate solution.
- P-toluidine reagent.
- 5% sodium arsenite solution ($NaAsO_2$).

2.2.22.1.2   *Procedure*

- **Visual standards**: Pipette into a 100-ml Nessler cylinder 1, 2, 3, ml etc. of dilute chromate-dichromate solution and dilute to 100 ml with phosphate. Protect the solutions from dust, evaporation, and direct sunlight. These standard solutions correspond to residual chlorine equivalent of 0.01, 0.02, 0.03 mg/l respectively.
- **Free available chlorine**: In a Nessler tube take 0.5 ml of toluidine reagent and 9.5 ml of sample. Mix quickly and thoroughly and add 0.5 ml of $NaAsO_2$ followed by mixing. Compare the color with standard (**call this reading A**).
- **Total available residual chlorine**: In a Nessler tube take 0.5 ml of toluidine reagent and 9.5 ml of sample, mix, and compare the color with standard color in similar tube after five minutes (**call this reading B**).
- **Blank**: Prepare a blank 9.5-ml sample and 0.5 ml of $NaAsO_2$ solution and compare the color with standard color immediately and after five minutes (**call this reading C1 and C2**).

2.2.22.1.3   *Calculation*

Free available residual chlorine = Reading A − Blank C1
Combined available residual chlorine = Reading B − Blank A
Total available residual chlorine = Reading B − Blank C2

Further Reading

APHA (1999). *Standard Methods for the Examination of Water and Wastewater*. Denver, CO: American Public Health Association, American Water Works Association and Water Environment Federation.

Maiti, S.K. (2004). *Handbook of Methods in Environment Studies: Water & Wastewater Analysis*, vol. 1. Rajasthan: ABD Publishers.

Metcalf & Eddy Inc (2014). *Wastewater Engineering: Treatment and Resource Recovery, 5*. New Delhi: Tata McGraw-Hill.

3

Wastewater Treatment Technologies

3.1 Overview of Wastewater Treatment Technologies

Wastewater treatment technologies have received steadily greater attention across the world in response to environmental, economic, and societal limitations increasingly posed by conventional wastewater systems. New wastewater treatment technologies incorporate natural processes and are designed with sustainability in mind, in contrast to energy-intensive and

Wastewater Treatment Technologies: Design Considerations, First Edition. Mritunjay Chaubey.
© 2021 John Wiley & Sons Ltd. Published 2021 by John Wiley & Sons Ltd.

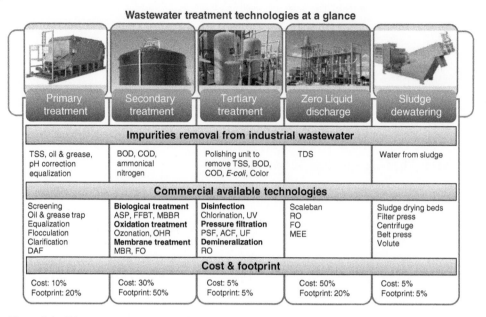

Figure 3.1 Wastewater treatment technologies at a glance.

chemical-dependent systems in current use. During the manufacturing process, wastewater is generated at various stages which is very complex in nature and highly variable in quantity and quality. Unless and until we adopt a structured approach toward collection, segregation, and treatment, it will be difficult to achieve the desired results. A host of new technologies and techniques for wastewater management are being developed around the world.

Sustainable Development Goal target 6.3 (improve water quality, wastewater treatment, and safe reuse) focuses mainly on collecting, segregating, treating, and reusing wastewater from households and industry, reducing diffuse pollution and improving water quality. According to the UN Sustainable Development Goal 6 Synthesis Report on Water and Sanitation, freshwater quality is at risk globally. Freshwater pollution is prevalent and increasing in many regions worldwide. Preliminary estimates of household wastewater flows, from 79 mostly high- and high-middle-income countries, show that 59% is safely treated. For these countries, it is further estimated that safe treatment levels of household wastewater flows with sewer connections and on-site facilities are 76 and 18%, respectively.

Wastewater treatment is generally a complex and multistage industrial process. Figure 3.1 shows wastewater treatment technologies at a glance. Most all-modern wastewater treatment plants consist of primary treatment, secondary treatment, tertiary treatment, zero liquid discharge (ZLD) (optional), and sludge dewatering units.

3.2 Primary Treatment

Primary treatment is a first-stage treatment of wastewater to remove large floating matter, oil and grease (O&G), grits, and suspended and colloidal particles from wastewater. Normally it consists of the following components:

- Screening.
- O&G trap.

- Equalization.
- Coagulation.
- Primary clarification.
- Dissolved air flotation (DAF).

3.2.1 Screening

A screen is a device with uniform openings for removing bigger suspended or floating matter in wastewater (see Figure 3.2). The dimensions of the channel and the size of the openings between the bars and their slopes are determined by conforming to the following hydraulic requirements and specifications:

- **Velocity**: The maximum velocity through the screen is 30 cm/s.
- **Slope**: Slope of the hand-cleaned screens should be between 30° and 45°. Horizontal and mechanically cleaned screens may have a slope between 45° and 80°.
- **Bar sizes and openings**: Usual bar size and openings of manually cleaned screens are shown in Table 3.1.

3.2.2 Oil and Grease Trap

An O&G trap is a specially designed rectangular tank to remove the floating oil and grease from wastewater. Generally it is installed between the screen chamber and equalization tank. Free O&G can be removed by means of a simple gravity unit (oil trap), which can be very useful in order to protect the downstream units, as it is easy to manage and has low operating costs. It consists of a tank that provides quiescent zones and is equipped with baffles that

Figure 3.2 A manually cleaned bar screen.

Table 3.1 Usual bar sizes and openings of manually cleaned bar screens.

Dimension of the bar facing flow (mm)	Clear spacing between bars (mm)	Area of opening (%)
6	18	75
6	24	80
6	30	83
6	36	85
9	18	67
9	24	73
9	30	77
9	36	80
12	18	60
12	24	61

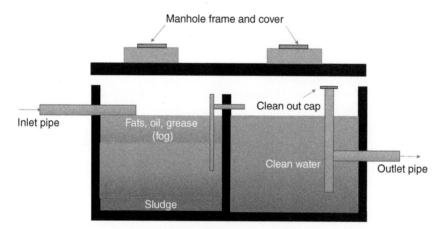

Figure 3.3 Schematic representation of a simple gravity O&G trap.

enhance oil/water separation by inducing proper direction to the flow (see Figure 3.3). Due to the usually high content of O&G in the wastewater to treat (ranging from 50 to 1500 ppm), the installation of an oil trap unit downstream of the screening phase is strongly recommended for preliminary free O&G removal to protect the secondary treatment units.

Moreover, depending on the amount of emulsified O&G in the wastewater, an enhanced DAF treatment with de-emulsifier dosage can be provided, to convert emulsified oil into free oil that can be separated by gravity thanks to flotation induced by air bubbling. A DAF treatment is considered to be particularly useful when high amounts of total suspended solids (TSS) and O&G are present (especially if with a high percentage of emulsified oils).

3.2.3 Equalization Tank

An equalization tank is a specially designed rectangular tank to make the effluent homogeneous. The tank also acts as a feedstock for the flocculation unit. The equalization unit will have the following functions:

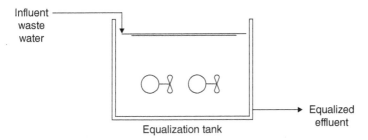

Figure 3.4 Schematic representation of an equalization tank.

- Reduction of possible peaks of hydraulic flow in the influent water.
- Reduction of possible load variations in the influent water.

The equalization unit protects the downstream biological process which is very sensitive to flow and quality variations. Equalization is a common practice for industrial wastewater treatment, especially if discontinuous flows are conveyed to the bioreactor, such as wastewater from membrane cleaning operations or dewatering operations. The equalization tank should be designed based on a minimum hydraulic retention time (calculated by the total flow) over 24 hours (see Figure 3.4).

3.2.4 Coagulation

Different effluent contains suspended matter and colloidal matter in varying proportions. A coagulant like ferric chloride, alum, or equivalent is dosed for coagulation purposes. Due to the effect of coagulant, heavy flocs are formed which trap the minute turbid particles and undissolved colloidal particles which otherwise would have passed through the clarification. Thus coagulants aid in better clarification. Dosage rate of coagulant is decided by jar testing in the field.

Polyelectrolyte helps in better flocculation due to its special characteristic of absorbing micron-sized particles to form a heavy floc. This is the latest innovation in the field of flocculation. The purpose of providing polyelectrolyte dosing after coagulant dosing is to achieve a better result to make effluent as clear as possible.

3.2.5 Primary Clarification

The primary clarification unit is a circular or rectangular bottom hopper tank fitted with a mechanical scraper to remove the floc formed in the coagulation and flocculation process.
A typical schematic diagram of a primary clarifier is shown in Figure 3.5.

3.2.6 Dissolved Air Flotation

The dissolved air flotation (DAF) unit (see Figure 3.6) is the preferred treatment to remove TSS and O&G (and eventually phosphorus), installed downstream of the equalization unit when O&G concentration in raw wastewater is higher than 100 mg/l. If surfactants are present in high concentration in the wastewater to be treated, DAF treatment should be

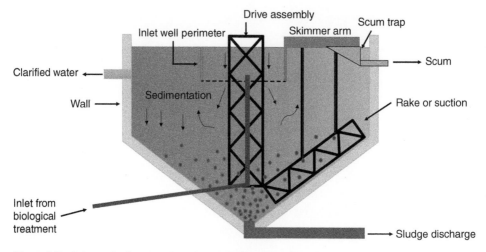

Figure 3.5 Schematic diagram of a primary clarification unit.

Figure 3.6 Dissolved air flotation (DAF).

avoided in order to prevent foam formation issues, and the clariflocculation system should be preferred.

The DAF unit performs the following operations:

- O&G, fats, and proteins removal (with de-emulsification of emulsified O&G).
- TSS removal (with efficiency determined based on the upstream TSS content).
- Chemical removal of phosphorus (when needed).

The DAF unit should include:

- A pH control system.
- Coagulation-flocculation tanks.
- Sedimentation and flotation basin.

- Air saturation unit.
- All the electromechanical and instrumentation and control (I&C) equipment (pumps, compressor, mixers, piping, valves, instruments, etc.) necessary for proper operation.

3.3   Secondary Treatment

Secondary treatment is a second-stage treatment of wastewater to achieve the following objectives:

- To remove or reduce the organic compounds concentration from wastewater.
- To convert (oxidize) dissolved and particulate biodegradable constituents into acceptable end products.
- To improve the concentration of dissolved oxygen (DO) in wastewater.

Secondary treatment of wastewater is classified into three categories:

- Biological treatment.
- Advanced oxidation treatment.
- Membrane treatment.

3.3.1   Biological Treatment

Biological wastewater treatment is a process that uses natural biological culture to decompose the organic substances present in wastewater. Normally it contains a biological reactor and secondary clarifier. Biological reactor degradation of organic matters takes place with the help of microorganisms, and in the secondary clarifier, settling of active microorganisms takes place, which are again recycled back into the biological reactor for further degradation. A schematic diagram of typical biological treatment is shown in Figure 3.7.

There are several methods available for the biological treatment of wastewater. These may be classified into two categories: aerobic and anerobic treatment. Both approaches

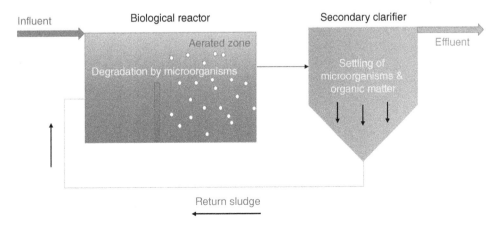

Figure 3.7   Schematic diagram of biological treatment.

essentially involve biochemical oxidation to remove organic contamination. Below the various biological aerobic as well as anaerobic treatment technologies are discussed briefly:

- **Aerobic treatment:**
  - Activated sludge process (ASP).
  - Fixed film bioreactor (FFBR).
  - Moving bed biofilm reactor (MBBR).
  - Trickling filter.
  - Oxidation pond.
  - Rotating biological contractor (RBC).
  - Root zone technology.
- **Anaerobic treatment:**
  - Up-flow anaerobic sludge blanket (UASB).
  - Packed anaerobic bed reactors (PABRs).
  - Hybrid anaerobic lagoons (HALs).
  - Packaged anaerobic systems.

3.3.1.1 Aerobic Treatment

Aerobic treatment is a method of wastewater treatment that uses oxygen to break down organic impurities and other pollutants like nitrogen and phosphorus. Oxygen is continuously mixed into the wastewater by a mechanical aeration device, such as an air blower or compressor. Aerobic microorganisms then feed on the wastewater's organic matter, converting it into carbon dioxide and biomass which can be removed.

Under aerobic treatment several technologies are used for secondary wastewater treatment. Some important aerobic secondary wastewater treatment technologies are ASP, fixed film bioreactor, MBBR, trickling filter, oxidation pond, RBC, root zone technology, and membrane bioreactor.

3.3.1.1.1 *Activated Sludge Process*

ASP is a biological treatment process of wastewater in which waste organic matter is aerated in an aeration tank with aerators and microorganisms which metabolize the soluble and suspended organic matter (see Figure 3.8). Part of the organic matter is synthesized into new cells and part is oxidized to carbon dioxide and water. The new cells formed in the reaction are removed from the liquid stream in the form of a flocculent sludge in the settling tank. Part of this activated sludge is recycled through the secondary clarifier into the aeration tank and the remaining material forms excess sludge.

Based on effluent characteristics, any of following units or a combination of all units may be selected:

- Activated sludge unit (ASU).
- Denitrification/nitrification biotreater (DNB).
- Intermittent nitrification/denitrification biotreater (INDB).

3.3.1.1.1.1 Principles of Aerobic Treatment

Aerobic treatment is a biological process that uses the capabilities of aerobic bacteria to remove organic pollutants and inorganic nutrients such as nitrogen and phosphorus.

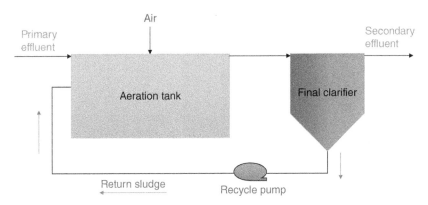

Figure 3.8 Schematic diagram of activated sludge process.

The term aerobic refers to the fact that bacteria and thus the aerobic process require oxygen (air). The following reactions occur:

Removal of organic pollutants, expressed as biological oxygen demand (BOD), chemical oxygen demand (COD), or total organic carbon (TOC):

$$BOD + O_2 + Biomass + nutrients \left(N/P\right) \rightarrow CO_2 + H_2O + new \ biomass + energy$$

Removal of nitrogen:

$$Nitrification: NH_4^+ + 1\tfrac{1}{2}O_2 \rightarrow NO_2^- + 2H^+ + H_2O \boxed{aerobic \ conditions}$$
$$NO_2^- + \tfrac{1}{2}O_2 \rightarrow NO_3^-$$

$$Denitrification: NO_3^- + BOD \rightarrow N_2 + H_2O + CO_2 \boxed{anoxic \ conditions}$$

The ASU is used for removal of organic pollutants from wastewater. In the DNB and INDB system, nitrogen is removed biologically by the nitrification/denitrification reactions. Note that phosphorus (P) can be removed both chemically (by precipitation) and biologically. However, P is normally absent in refinery effluent. Removal is therefore not required and is not considered further in this section. To have high biological conversion rates, sufficient sludge (i.e. microorganisms) must be present in the biological system. In aerobic systems this is accomplished by decoupling sludge residence time from hydraulic residence time. The most common method is to recover the sludge leaving the biotreater by means of sedimentation in a clarifier and recycle the sludge back to the reactor. This enables an aerobic biotreater to run at much shorter hydraulic retention times than would be possible if the sludge was not recovered. Furthermore, slow-growing organisms needed for removal of recalcitrant BOD and for nitrification can than be maintained within a unit operating at relatively short hydraulic retention times.

3.3.1.1.1.2 Description of Aerobic Treatment Systems
In the following a scheme and short description of the ASU, DNB, and INDB are given.

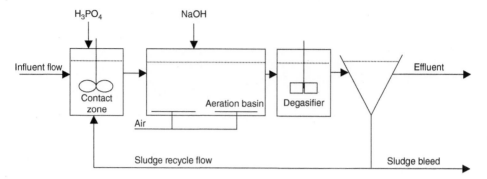

Figure 3.9 Schematic diagram of activated sludge unit.

3.3.1.1.1.2.1 Activated Sludge Unit (ASU)   In the ASU, organic pollutants (BOD) are removed biologically by oxidizing the organic matter into CO_2, H_2O, and new biomass (sludge). Dependent on the ammonia and total Kjeldahl nitrogen demand, oxidation of ammonia to nitrate occurs as well. The ASU consists of the following main processing units (see Figure 3.9):

- **Contact zone**: In the contact zone the incoming influent is mixed with recycled sludge. The selector is equipped with a mechanical mixer. The function of the contact zone is to favor conditions for the development of good settling sludge by promoting floc-forming organisms over filamentous organisms. The latter are known to cause a bad settling bulking sludge.
- **Aeration basin**: In the aeration basin organic pollutants (BOD, COD) are biologically oxidized to CO_2, H_2O, and new sludge using O_2. Oxidation of sulfide (to sulfate) and/or ammonia (to nitrate), called nitrification, may occur as well. The aeration basin is continuously aerated and mixed so that O_2 is supplied and the content is kept homogenous. Aeration/mixing is done by aeration equipment (usually bubble aeration). Mechanical mixing is normally not required.
- **Degasifier**: In the degasifier air bubbles attached to the sludge are released and the sludge is allowed to flocculate. A slow paddle mixer supports these processes.
- **Clarifier**: The sludge water mixture enters the clarifier where sludge and water are separated by sedimentation. The settled sludge is recycled; a small amount is discharged. The treated water, virtually free from solids, is discharged or further treated.

In the ASU settled sludge is continuously recycled to the selector and hence the aeration basin. Sludge recycling is needed to maintain the sludge concentration in the aeration basin at the desired level. A small amount of sludge, compensating for the sludge production in the system, is discharged as sludge bleed. Chemical dosing includes H_3PO_4 (P-source) and/or NaOH (pH control). Optionally polyelectrolyte is dosed into the degasifier to support sludge settling.

3.3.1.1.1.3 Denitrification/Nitrification Biotreater
In the DNB organic pollutants (BOD) are removed biologically by oxidizing the organic matter into CO_2, H_2O, and new biomass (sludge). In addition, nitrogen is removed by

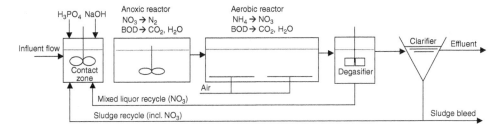

Figure 3.10 Schematic diagram of denitrification/nitrification biotreater unit.

nitrification/denitrification. Nitrification and denitrification is carried out in separate reactors. The DNB consists of the following main processing units:

- **Contact zone**: In the contact zone the incoming influent is mixed with recycled sludge and recycled mixed liquor. The contact zone is equipped with a mechanical mixer. The function of the selector is as in the ASU.
- **Anoxic basin**: In the anoxic basin nitrate is denitrified to N_2 (denitrification) and BOD is oxidized to CO_2, H_2O, and new sludge. BOD is oxidized with NO_3 instead of O_2. The anoxic tank is equipped with mechanical mixers.
- **Aeration basin**: In the aeration basin BOD not removed in the anoxic tank is oxidized to CO_2, H_2O, and new sludge using O_2. In addition, ammonia is oxidized to nitrate (nitrification). For mixing/aeration, aerators (usually bubble aeration) are used. Mechanical mixing is normally not required.
- **Degasifier**: In the degasifier air bubbles attached to the sludge are released and the sludge is allowed to flocculate. A slow paddle mixer supports these processes.
- **Clarifier**: The sludge water mixture enters the clarifier where sludge and water are separated by sedimentation. The settled sludge is recycled; a small amount is discharged. The treated water, virtually free from solids, is discharged or further treated (see Figure 3.10).

In the DNB settled sludge is continuously recycled to the selector and hence the anoxic and aeration basin. Sludge recycling is needed to maintain the sludge concentration in the aeration basin at the desired level. A small amount of sludge, compensating for the sludge production in the system, is discharged as sludge bleed. In the DNB nitrate is continuously recycled from the degasifier to the selector and hence the anoxic tank. This recycling is needed to supply the anoxic tank with sufficient nitrate for the denitrification process. The mixed liquor recycling sets the efficiency of denitrification and the nitrogen removal. Chemical dosing includes H_3PO_4 (P source), NaOH (pH control), or methanol (COD source at low COD/N ratio). Optionally PE is dosed to the degasifier to support sludge settling.

3.3.1.1.1.4 Intermittent Nitrification/Denitrification Biotreater

In the INDB organic pollutants (BOD) are removed biologically by oxidizing the organic matter into CO_2, H_2O, and new biomass (sludge). In addition, nitrogen is removed by nitrification/denitrification. Nitrification and denitrification is carried out in one reactor. The INDB consists of the following main processing units (see Figure 3.11):

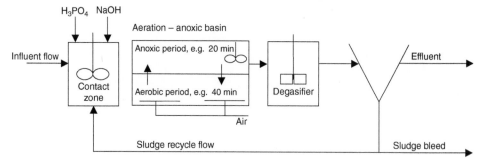

Figure 3.11 Schematic diagram of intermittent nitrification/denitrification biotreater unit.

- **Contact zone**: In the contact zone the incoming influent is mixed with recycled sludge. The selector is equipped with a mechanical mixer. The function of the contact zone is to favor conditions for the development of good settling sludge by promoting floc-forming organisms over filamentous organisms. The latter are known to cause a bad settling bulking sludge.
- **Aeration/anoxic basin**: In this basin BOD is oxidized to CO_2, H_2O, and new sludge using O_2 and/or NO_3. In addition, ammonia is oxidized to nitrate (nitrification) and denitrified to N_2. To allow for nitrification and denitrification in one reactor the system is alternately aerated and not aerated but mixed. For aeration and mixing the basin is equipped with an aerator (usually bubble aeration) and mechanical mixers.
- **Degasifier**: In the degasifier air bubbles attached to the sludge are released and the sludge is allowed to flocculate. A slow paddle mixer supports these processes.
- **Clarifier**: The sludge water mixture enters the clarifier where sludge and water are separated by sedimentation. The settled sludge is recycled; a small amount is discharged. The treated water, virtually free from solids, is discharged or further treated.

As in the ASU system sludge is recycled and discharged. A mixed liquor recycle as in the DNB system is not needed. Chemical dosing includes H_3PO_4 (P source), NaOH (pH control), or methanol (COD source at low COD/N ratio). Optionally, poly electrolyte is dosed to the degasifier to promote sludge settling.

3.3.1.1.2 Fixed Film Bioreactor

Fixed film bioreactor technology involves aerobic biological wastewater treatment based on the attached growth principle. Natural microbiological metabolism in the aquatic environment is capitalized in this process. Under proper environmental conditions the soluble organic substances of the waste are completely destroyed by biological oxidation. In the bioreactor, part of the organic substances are oxidized and the rest converted into biological mass. The end products of the metabolism are gas, liquid, and the synthesized biological mass which can flocculate easily and be separated out in a clarifier [1]. A schematic diagram of a fixed film bioreactor plant is shown in Figure 3.12.

Wastewater enters the top of the bioreactor packed with corrugated PVC fills, where it is aerated. The bacteria grow and attach themselves to the surface of packing materials which helps minimize the cell washout. The packing breaks the diffused air stream into fine

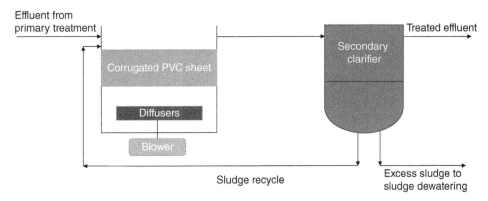

Figure 3.12 Schematic diagram of fixed film bioreactor plant.

bubbles, resulting in oxygen transfer so efficient that the system uses only about half the air of conventional systems.

3.3.1.1.3 Moving Bed Biofilm Reactor

MBBR is a combination of suspended growth and attached growth process using the whole tank volume for biomass growth. It uses simple floating media, which are carriers for attached growth of biofilms. Biofilm carrier movement is caused by the agitation of air bubbles. This compact treatment system is effective in removal of organic matter as well as nitrogen and phosphorus, while facilitating effective solids separation. An MBBR technology process flow diagram is shown in Figure 3.13.

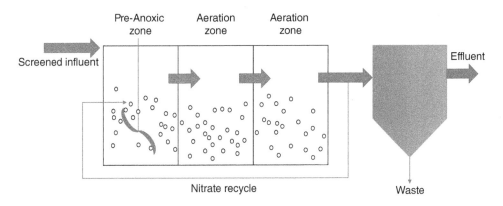

Figure 3.13 MBBR technology process flow diagram.

3.3.1.1.4 Trickling Filter

A trickling filter is a non-submerged fixed film biological reactor using rock or plastic packing over which wastewater is distributed continuously. Treatment occurs as the liquid flows over the attached biofilm. The depth of the rock packing ranges from 0.9 to 2.5 m and averages 1.8 m. Rock filter beds are usually circular and the liquid wastewater is distributed over the top of the bed by a rotary distributor. Many conventional trickling filters using rock as the packing material have been converted to plastic packing to increase the treatment efficiency of trickling filters [2].

3.3.1.1.4.1 Advantages
- Simplicity of operation.
- Resistance to shock loads.
- Low sludge yield.
- Low power requirements.

3.3.1.1.4.2 Disadvantages
- Relatively low BOD removal (85%).
- High suspended solids in the effluent (20–30 mg/l).
- Little operational control.

3.3.1.1.5 Oxidation Pond

Oxidation ponds are an effective, low-cost, and simple technology for reducing the BOD of wastewater. Small ponds that receive a reasonably high input of plant nutrients generally develop ecosystems that feature algal populations that produce oxygen in excess of the respiration requirements of the algae. This "excess" oxygen can be used by bacteria to oxidize biodegradable organic matter entering the pond. In natural-aeration oxidation ponds, algal production of oxygen occurs near the surface of ponds, in a "euphotic" zone limited by the depth of light penetration into the pond. A small amount of oxygen also enters the pond by surface diffusion. If the depth of a pond is greater than the euphotic zone, and wind mixing does not mix the near-surface oxygenated water the full depth of the pond, or if the amount of oxygen produced or transferred to deeper water is less than is needed by bacterial respiration there, then an anaerobic (no oxygen) zone develops at the bottom of the pond. How far the anaerobic zone extends toward the pond surface depends on the balance of oxygen supply and oxygen demand. Oxygen demand in a pond depends on the loading of biodegradable organic waste to the pond.

The concepts outlined above are used in the design of oxidation ponds. There are fundamentally three types of pond, depending on the dissolved oxygen (DO) profile in the pond. At one extreme are shallow (approx. 0.5 m deep), fully aerobic ponds, where a euphotic zone that reaches the bottom and effective mixing cause DO to be present the full depth of the pond. Although the BOD_5 (five days BOD) of the influent waste is very efficiently reduced, the concentration of algal and bacterial suspended solids in the effluent can be high, resulting in a high BOD_5 concentration in the effluent.

At the other extreme are anaerobic ponds, where organic loading is so high relative to oxygen entering the pond that the pond is anaerobic right to the surface. Under these conditions, fermentation processes and anaerobic oxidation can remove about 70% of the BOD_5 of the waste. Anaerobic ponds are mixed to some extent by the bubbling of gases (carbon dioxide and methane) produced in the pond. The suspended solids turbidity caused by mixing reduces light penetration, and algal production of oxygen, when it occurs at all in these ponds, is negligible.

Anerobic ponds are a very cost-effective way of reducing the BOD_5 of medium- to high-strength wastes. Anerobic ponds can be deep and thereby economical in terms of land use, and significant BOD_5 reduction is achieved without the need of energy input for treatment.

A third type of oxidation pond is the "facultative pond." The depth of natural-aeration facultative ponds – usually 1.0–1.5 m – is too deep for oxygen to penetrate to the bottom of the pond, and an anerobic zone develops there. Solids from the incoming waste settle into the anerobic sludge near the bottom of the pond and degrade anerobically, releasing soluble degradable organic material and nutrients which diffuse upwards in the pond. Near the top of the pond oxygen is supplied by algal photosynthesis and to a limited extent by diffusion from the air. There is DO present to only a few centimeters depth at night, but DO diffuses deeper during daylight. Thus there exists a fully aerobic zone at the top of the pond, and between this and the anerobic zone at the bottom there is a middle zone where oxygen is cyclically present and bacterial respiration is "facultatively" aerobic–anerobic.

3.3.1.1.5.1 Advantages
- Low operational and maintenance cost.
- Lagoons provide effective treatment with minimal threat to the environment.
- Work well in clay soils where conventional subsurface onsite absorption fields will not work.

3.3.1.1.5.2 Disadvantages
- Lagoons must be constructed in clay soil or be lined to prevent leakage.
- May overflow occasionally during extended periods of heavy rainfall.
- If there are extended periods of overcast windless days, offensive odors may occur for a brief time.
- Lagoons usually recover rapidly if this occurs.
- Cannot be installed on a small lot.
- Takes up a relatively large space for only one use.
- Lagoons are not esthetically acceptable to some people.
- Some people consider lagoons unsightly and unsafe. As with any other open body of water, there is some potential danger. Although lagoons are required to be fenced, this does not always prevent access by people or pets. They should be moved on a regular basis during the growing season.
- No trees should be allowed to grow around the lagoon.
- A fence should be constructed to discourage entry and control access.

3.3.1.1.6 *Rotating Biological Contractor*
An RBC consists of a series of closely spaced circular disks of polystyrene or polyvinyl chloride that are submerged in wastewater and rotated through it. The cylindrical plastic disks are attached to a horizontal shaft and are provided at standard unit sizes of approximately 3.5 m in diameter and 7.5 m in length [2]. The surface area of the disks for a standard unit is about 9300 m^2, and a unit with a higher density of disks is also available with approximately 13 900 m^2 of surface area. The RBC unit is partially submerged (typically 40%) in a tank containing the wastewater, and the disks rotate slowly at about 1.0–1.6 rpm. As the RBC disks rotate out of the wastewater, aeration is

accomplished by exposure to the atmosphere. Wastewater flows down through the disks, and solids sloughing occurs.

3.3.1.1.6.1 Advantages
- Simplicity of operation.
- Resistance to shock loads.
- Low sludge yield.
- Low power requirements.

3.3.1.1.6.2 Disadvantages
- Relatively low BOD removal.
- High suspended solids in the effluent.
- Little operational control.
- Bad smelling problem.
- Large space requirement.

3.3.1.1.7 Duckweed Pond Technology
Duckweed pond technology is a treatment method of wastewater in which duckweed plants. The duckweed pond is an earthen basin, preferably lined, where duckweed plants grow and cover the entire water surface. Since the floating aquatic plant is very small (duckweed is only a few millimeters in size) it can be easily swept off from the water surface to one side by wind or waves. Therefore a floating grid system of plastic, bamboo, or any other suitable material has to be installed to ensure quiescent conditions allowing duckweed to maintain a uniform cover over the entire pond surface. The excess duckweed biomass is regularly harvested. Fresh duckweed can be introduced into separate fishponds to grow fishes. The duckweed can also be dried or converted into pellets for use as feed for chicken and cattle.

3.3.1.1.8 Root Zone Technology
Root zone technology is an artificially constructed wetland system in which wastewater is kept at or above the soil surface and appropriate vegetation for sufficient time. The three essential components of the system include the soil, the appropriate vegetation such as reeds, cattails, bulrushes, and sedges, and the microbial organisms. The system has been used for the treatment of industrial wastewater. COD reductions of 84% have been reported from the textile plant effluent with COD around 1500 ppm at hydraulic residence time of 28 days [3].

3.3.1.2 Anerobic Treatment
Anerobic wastewater treatment is a biological process where microorganisms degrade organic contaminants in the absence of oxygen. The bioreactor contains a thick, semi-solid substance known as sludge, which is comprised of anerobic bacteria and other microorganisms. These anerobic microorganisms, or "anaerobes," digest the biodegradable matter present in the wastewater, resulting in an effluent with lower BOD, COD, and/or TSS, as well as biogas byproducts.

Under anerobic treatment several technologies are used for secondary wastewater treatment. Some important anerobic secondary wastewater treatment technologies are UASB reactors, PABRs, HALs, and packaged anerobic systems.

3.3.1.2.1 Up-Flow Anerobic Sludge Blanket Reactors

UASB reactors are suitable for low- to medium-strength wastewater. This is anaerobic treatment of wastewater in which effluent is pumped to the reactor and dispersed through a unique distribution system that ensures uniform distribution of effluent throughout the sludge blanket. The microorganisms degrade organic impurities and generate biogas. A three-phase separator located in the upper portion of the reactor separates gas, liquid, and sludge fractions. Biogas is collected in domes and transported, while suspended solids settle back into the sludge blanket, retaining valuable bacterial population. Liquid overflows the effluent gutter for further treatment. A schematic diagram of a modern UASB is shown in Figure 3.14.

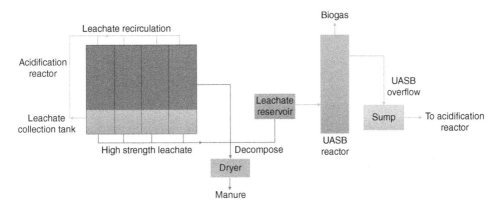

Figure 3.14 Schematic diagram of a modern UASB reactor.

3.3.1.2.2 Packed Anerobic Bed Reactors

PABRs are designed for providing treatment for high-strength, low-solids wastewaters. PABRs are hybrid systems consisting of both suspended-growth and fixed-film sections, which makes these systems capable of handling relatively high organic and hydraulic loads at short retention times.

3.3.1.2.3 Hybrid Anerobic Lagoons

HALs are suspended-growth systems designed to provide treatment of very high-strength, high-solids wastewaters. They feature advanced mixing and flow management to promote high removal efficiencies, even under significant load variations. HALs are often designed to handle high-solids wastewaters that may otherwise require physical/chemical pretreatment prior to a biological process.

3.3.1.2.4 Packaged Anerobic Systems

These modular anaerobic systems, available in suspended-growth or fixed-film configurations, are suited for small wastewater flows of 2000–20 000 gallons per day. These units can be delivered complete to a project site, requiring only simple electrical and piping connections. The modular design allows for easy expansion and movement to new locations.

3.3.2   Advanced Oxidation Treatment

Advanced oxidation treatment is the partial or complete conversion of organic compounds into carbon dioxide and water. When only partial conversion occurs, portions of the incompletely destroyed organic compounds may be converted to partially oxidized forms of the original compounds such as alcohols, aldehydes, and ketones, or carboxylic acids, which are more biodegradable than the parent compounds. Common oxidation methods include ozone, chlorine, chlorine dioxide, peroxide, wet air, supercritical water, and potassium permanganate. One common class of advanced oxidation, employs the use of generated OH^- radicals to achieve conversion of organic compounds. Some common chemical oxidation processes, for example, chlorination and ozonation, are commonly used for water disinfection. Another method to perform disinfection is by means of ultraviolet (UV) systems, which involve the use of lamps producing UV light which destroys microorganisms' genetic material. The effectiveness of UV radiation depends on the dose received by the microorganisms, path length from source to microorganisms, contact time, and water turbidity. Differently from other disinfection treatments, UV disinfection does not form toxic byproducts, but at the same time does not guarantee a disinfection residual activity in water after the treatment is performed.

There are several methods available for advanced oxidation treatment of wastewater. Below the various technologies are discussed briefly.

3.3.2.1   Ozonation

Ozone was discovered in the nineteenth century, generally produced during lightning storm. Ozone is a tri-atomic allotrope of oxygen, having a molecular weight of 48, which is one of the most powerful oxidants. It is blue gas with pungent odor with a concentration of approximately 0.01 ppm and is easily detectable in air by most people. It is 1.5 times more dense than O_2 and 12.5 times more soluble in water. Ozone has a strong absorption band in infrared visible and ultraviolet radiation.

Ozonation is a chemical water treatment technique based on the infusion of ozone into water and wastewater. Ozonation is a type of advanced oxidation process (AOP), involving the production of very reactive oxygen species able to attack many organic compounds and all microorganisms. The treatment of wastewater with ozone has a wide range of applications, as it is efficient for disinfection as well as for the degradation of organic and inorganic pollutants. Ozone is produced with the use of energy by subjecting oxygen (O_2) to high electric voltage or to UV radiation. The required amounts of ozone can be produced at the point of use.

Ozone can be effectively used for the treatment of sewage and industrial wastewater. The application may be pre ozonation or post ozonation. Ozone is a powerful oxidant, leaves no residual harmful product, has no sludge disposal problem, and increases the DO content of wastewater which helps further in the degradation of residual pollutant. Therefore ozone has a use in the treatment of all types of wastewater, such as sewage wastewater and various industrial wastewaters in the chemical, paper, and textile industries. It is also used for the removal of color, cyanide, and phenolic compounds from wastewater. All these properties have made ozone nowadays an ultimate treatment for all types of wastewater treatment.

The combination of ozone oxidation followed by biological treatment has been installed full-scale at a large German industrial chemical complex. The integrated process (ozone–aerobic oxidation–ozone) achieved more COD reduction than the non-ozonated process over a similar treatment period. Some experimental data found that biodegradability, expressed as BOD_5/COD ratio, increased from zero up to a maximum value of 0.21, corresponding to the ozone consumption of 0.7 mg/mg of reduced COD.

3.3.2.1.1 Ozone Generation

Because of its relatively short half-life, ozone is generated onsite by an ozone generator. The conventional ways to produce ozone are UV-light and corona-discharge. Ozone generation by corona-discharge is most common nowadays and has many advantages, such as longer lifespan of the unit, higher ozone production, and higher cost efficiency. Production with UV-light is an option where only small amounts of ozone are required. Other ozone generators that are available involve electrolysis of water and the use of membranes. With this method, ozone is dissolved in the process water as soon as it is formed, resulting in ozonation using minimum equipment. A typical ozonation system consists of an ozone generator and a reactor where ozone is bubbled into the wastewater to be treated.

3.3.2.1.1.1 Operation and Maintenance

Ozone generation uses a significant amount of electrical power. Constant attention must thus be given to the system to ensure that power is available. Moreover, ozone should not be released from the system and connections in or surrounding the ozone generator should not be leaking. The operator must monitor the appropriate subunits on a regular basis to ensure they are not overheating. Therefore the operator must check for leaks routinely, since a very small leak can cause unacceptable ambient ozone concentrations.

3.3.2.2 OH Radicals

Hydroxyl (OH) radicals are generated by an OH radical generator. Such radicals are harmless to the human body yet capable of sterilizing and disinfecting contaminants in wastewater. They are an excellent oxidizing agent. By reacting with any organic materials, they can kill various viruses and bacteria such as *Escherichia coli* O157, and salmonella. They also decompose all heavy metals, agricultural chemicals, and chemical components.

OH radicals are very powerful oxidizers and work very well even on higher TDS (TDS < 100 000 ppm). They can be mixed in wastewater with the help of microbubble nano-technology. Microbubble nano-generators generate <5 μm bubbles inside wastewater. OH radicals being powerful oxidizers once get mixed with effluent, the organics inside wastewater get oxidized, and finally very good COD reduction takes place inside wastewater. The working principles of OH radical technology are shown in Figure 3.15.

3.3.2.2.1.1 Advantages

- OH radicals have germicidal power about 3600 times that of chlorine, 2000 times ozone, and 180 times UV rays. They react to any organic materials to sterilize and disassemble them.
- A natural material that is harmless to the human body.
- Decomposes to get rid of all heavy metal, agricultural chemicals, and chemical components.

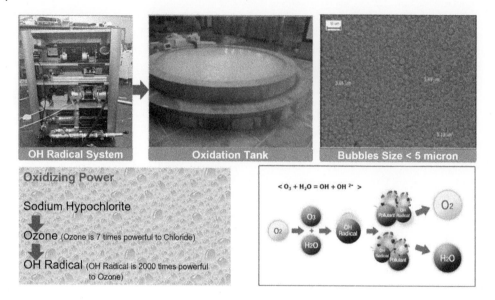

Figure 3.15 Working principle of OH radical technology.

- Other than possessing qualities of germicide and disinfectant, they also remove odor from waste water, serving to improve overall water quality.
- Increase the quantities of DO, anion, and minerals.
- Since they do not use filter or chemicals, maintenance cost is zero.

3.3.2.3 Solar Detoxification

The removal of toxic elements and organic compounds from wastewater using catalyst in the presence of UV radiation may be referred to as solar detoxification. Recently, it has been demonstrated that solar detoxification has great potential for the elimination of toxic elements, organic compounds, and biological contamination in wastewater. Solar detoxification is a process of treatment of wastewater in which titanium dioxide (TiO_2) is exposed to the sun, the catalyst absorbs the high-energy photon light from the UV portion of the solar spectrum, and reactive chemicals known as hydroxyl radicals are formed. These radicals are powerful oxidizers and disinfectants. A concentration of 0.01% TiO_2 is most effective in killing bacteria.

The configuration of solar detoxification systems, used for removal of toxic elements, organic compounds, and biological contamination at the tertiary stage of the treatment sequence of wastewater, is shown in Figure 3.16. The system consists of a shallow solar pond reactor made of concrete material for holding the effluent. Since industries already use holding ponds for microbiological treatment of wastewater, the same shallow ponds can be modified as reactors. The reactor should be fitted with a diffused aerator for catalyst agitation. The solar detoxification reactor can be operated in a slurry configuration. The catalyst TiO_2 at concentration 0.1% should be mixed with effluent in the reactor. The other important component of this system is UV radiation. During day-time natural available solar radiation can be utilized for these purposes, and during night or rainy days UV

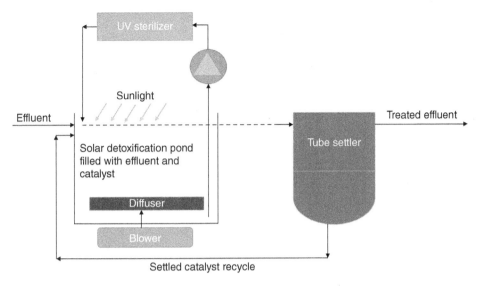

Figure 3.16 System configuration of solar detoxification process.

sterilizer can be used. After mixing the effluent with catalyst, and allowing for a residence time of 50 minutes, the effluent should be passed through a specially designed high-rate tube settler for clarification of catalyst. The settled catalyst should be recycled to reuse in the solar detoxification reactor. More details on solar detoxification technology can be found in Chapter 5.

3.3.2.4 Electro-oxidation

EO is a technique used for wastewater treatment, mainly for industrial effluents, and is a type of AOP. AOPs have shown to be very useful technologies for application in different wastewater treatment sectors. These processes use the very strong oxidizing power of hydroxyl radicals to oxidize organic compounds to carbon dioxide and water. These procedures usually involve the use of O_3, H_2O_2, Fenton's reagent, and electrolysis to generate the hydroxyl radicals.

The EO unit comprises two electrodes, operating as anode and cathode, connected to a power source. When an energy input and sufficient supporting electrolyte are provided to the system, strong oxidizing species are formed, notably hydroxyl radicals at anode. To assure a reasonable rate of generation of radicals, voltage is adjusted to provide current density of approx. $10–100\,mA/cm^2$. The cathode materials are mostly the same in all cases; the anodes can vary greatly according to the type of application, as the reaction mechanism is strongly influenced by the anode material selection. Cathodes are mostly made up of stainless-steel plates, platinum mesh, or carbon felt electrodes. Anode material can be made up of carbon and graphite, platinum, m (MMOs), lead dioxide, or boron-doped diamond (BDD), depending on application. These hydroxyl radicals are highly reactive oxidants that can react with nearly all organic contaminants of wastewater and eventually mineralize them to CO_2 and H_2O non-selectively in ambient pressure and atmospheric temperature. The refractory organic compounds of wastewater are thus converted into reaction intermediates and, ultimately, into water and CO_2 by complete mineralization.

EO is the latest technology for treating harmful and recalcitrant organic pollutants, which are typically difficult to degrade with conventional wastewater treatment technologies. In EO there is no requirement of any external chemicals. The system has been applied to treat a wide range of harmful and non-biodegradable contaminants, including aromatics, pesticides, drugs, and dyes. In some industrial applications EO is used as a final treatment technology for wastewater treatment. Due to its relatively high operating costs, it is often combined with other technologies, such as biological processes and ozonation. However, a recent study has found that the use of a coupled process using O_3 and EO increases the effectiveness of the process and also could reduce the operating costs associated with the application of AOPs. Another study reported that EO and ozonation provide a sustainable wastewater treatment process.

3.3.3 Advanced Membrane Treatment

The advanced membrane treatment process mainly depends on three basic principles, namely adsorption, sieving, and electrostatic phenomenon. The adsorption mechanism in the membrane separation process is based on hydrophobic interactions of the membrane and solute. These interactions normally lead to more rejection because it causes a decrease in the pore size of the membrane. The separation of materials through the membrane depends on pore and molecule size. For this reason, various membrane processes with different separation mechanisms have been developed. There are several methods available for the advanced membrane treatment of wastewater.

3.3.3.1 Membrane Biofilm Reactor

MBR technology for wastewater treatment involves hybridization of suspended growth biological treatment with ultrafiltration (UF) membranes. The membranes are used to perform the critical solid/liquid separation function. This is an effective technology for industrial wastewater treatment and water reuse due to its high product water quality and low footprint. Due to their robustness and flexibility, submerged MBR systems are increasingly preferred.

In an MBR, the membrane zone can be described as the initial step in a biological process where microbes are used to degrade pollutants which are then filtered by a series of submerged membranes. The individual membranes are housed in a tank. Air is introduced through diffusers at the bottom of the tank to continually scour membrane surfaces during filtration, facilitate mixing, and, in some cases, contribute oxygen to the biological process.

The MBR unit performs two unitary operations:

- A biological reaction, performed in the anoxic and oxidation treatment tanks.
- Solid/liquid separation achieved through UF membranes submerged in an external aerated reactor, which will perform a further oxidation process or through side-stream vertical membranes with an air-lift system.

The MBR should include:

- An aerated bioreactor provided with an initial anerobic selector.
- A membranes tank.

- A back pulse tank.
- Chemical storage and dosing systems for membrane cleaning.
- All the electromechanical and I&C equipment (pumps, compressor, piping, valves, mixers, instruments, etc.) necessary for proper operation.

The benefits of MBR include a reduced footprint – usually 30–50% smaller than an equivalent conventional active sludge facility with secondary clarifiers and pressure sand filtration.

3.3.3.2 Forward Osmosis

FO is a natural process and an integral part of the survival of flora and fauna on this planet. It is this process only that makes plants transport water from their root systems to their leaves and it provides the primary means of transporting water in cells across most organisms in nature. Both FO and the conventional reverse osmosis (RO) process are highly selective for water molecules. The difference lies in the means by which water molecules are driven through the membrane. The FO process is governed by a difference in osmotic pressure and the direction of water diffusion takes place from lower concentration (the feed side) to higher concentration (the draw side). RO processes, on the other hand, are governed by hydraulic pressure differences and the direction of water diffusion is from high concentration to low concentration. The FO process employs semipermeable membranes to concentrate the dissolved contaminants and separate fresh permeate from dissolved solutes. The driving force for this separation is an osmotic pressure gradient which is generated by a draw solution of high concentration to induce a net flow of water through the membrane into the draw solution, thus effectively separating the feed water from its solute. As osmosis is a natural phenomenon, it significantly requires less energy compared to the conventional RO process.

3.3.3.2.1.1 Advantages

- Can be used for highly saline waters which are impossible to treat through conventional RO process.
- Concentration of total dissolved solids (TDS) can be increased to the tune of 16–17%.
- Since natural osmosis process is used, the power consumption is relatively less compared to other conventional processes.
- Operation and handling are much easier and reliable.

More details on FO technology are provided in Chapter 5.

3.4 Tertiary Treatment

Tertiary treatment is a third-stage treatment of wastewater to remove pathogenic bacteria, small and micro suspended solids, color, and TDS from wastewater. Normally it consists of the following processes:

- Disinfection.
- Pressure filtration.
- Demineralization.

3.4.1 Disinfection

Chlorine (Cl_2) has been used for many years to treat municipal and industrial water and wastewater as a disinfectant, because of its capacity to kill most pathogenic microorganisms quickly. The effectiveness of chlorine is dependent on the chlorine concentration, time of exposure, and pH of the wastewater.

Chlorine as disinfectant is usually applied for the following purposes:

- To kill microorganisms: bacteria, algae, fungi, viruses, and higher organisms.
- To remove small amount of BOD and COD from wastewater.
- To remove small amounts of iron and heavy metals from wastewater.

For chlorination either chlorine gas or sodium hypochlorite (NaOCl) or calcium hypochlorite ($Ca(OCl)_2$) is added in wastewater storage tanks and a reaction time of 20–30 minutes should be allowed.

Disinfection treatment by means of sodium hypochlorite dosage is recommended downstream of biological treatment to remove any residual biological activity in order to prevent biofouling if sand or membrane filtration is installed downstream. Sodium hypochlorite is generally considered one of the safer chlorination agents, and is liquid and more easily manageable than other chlorine-gas products. It also shows oxidizing potential and can perform partial removal of organics as a secondary effect. The reaction takes place in a chlorination tank for a proper contact time installed downstream of the biological treatment.

Chlorination is an extremely effective, well-established method of disinfection. The ease with which dosing can be controlled makes it an appealing method, particularly for small to medium facilities with variable flow.

Other systems that can be considered for disinfection are UV lamps and ozonation. Ozone is a very effective disinfection alternative but is generated by applying high voltage to either air or pure oxygen, thus imposing an important energy requirement on the facility, which makes it less applicable. UV lamps are also effective in disinfecting wastewater but are not recommended for effluent with residual BOD or suspended solids.

3.4.1.1 Chlorination Chemistry
3.4.1.1.1 *Chlorine Gas*
The reaction of chlorine gas with wastewater is:

$$Cl_2 + \underset{H_2O}{\rightarrow} HOCl + HCl$$

HOCl is a weak acid whose undissociated form is 20–50 times more effective for killing microorganisms than is its ion OCl.

3.4.1.1.2 *Sodium Hypochlorite*
The reaction of sodium hypochlorite (NaOCl) with wastewater is:

$$NaOCl + \underset{H_2O}{\rightarrow} HOCl + NaOH$$

3.4.1.1.3 Calcium Hypochlorite

The reaction of calcium hypochlorite ($Ca(OCl)_2$) with wastewater is:

$$Ca(OCl)_2 + 2 \underset{H_2O}{\rightarrow} 2HOCl + Ca(OH)_2$$

3.4.2 Pressure Filtration

Pressure filtration is a method of filtration in which raw chlorinated water is pumped into a specially designed pressure vessel where refractory organics, colorants, and suspended solids can be removed. In pressure filtration, pressure sand filter (PSF) or multigrade filter (MGF) or dual media filter (DMF) can be used. UF can be also used under filtration.

For the design of vessels the following details should be noted, as shown in Table 3.2.

Table 3.2 Design parameters of various types of pressure filters.

Description	Surface loading ($m^3/h/m^2$)	Suspended solids (ppm)	Height of shell (m)	Media used	Media height (mm)	Density (kg/m^3)
PSF	7–15	01–30	1.5–2.0	Sand	1200	1780
MGF	9–18	30–70	2.0–2.5	Sand	700	1780
				Gravel	700	1780
DMF	9–18	30–70	2.0–2.5	Sand	350	1780
				Gravel	350	1780
				Anthracite	600	0900

Source: The above data are compiled by the author through various field experiences.

3.4.2.1 Ultrafiltration

UF is a membrane treatment that allows for a high performance of solids removal and also soluble organics removal, separating them from treated water particles up to 0.01 μm (see Figure 3.17). UF membranes also perform disinfection of the permeate, retaining almost completely all bacteria and partially viruses.

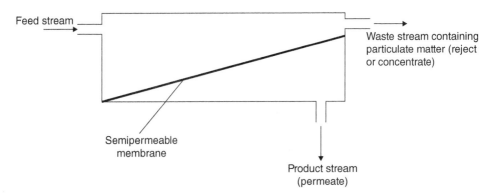

Figure 3.17 Schematic diagram of separation process through semipermeable membrane.

It can be suitably combined with aerobic biological treatment in the MBR process: within the MBR, UF membranes provide for solid/liquid separation process, typically producing an effluent of high quality in terms of suspended solids that can be fed directly to the desalination unit. This integration also allows for a significant saving in terms of plant footprint, as conventional large secondary clarifiers are replaced by membrane modules. If no MBR is installed in the biological section, the polishing section should consider PSF treatment followed by UF treatment as a pretreatment for the following desalination unit.

3.4.3 Demineralization

Demineralization is a tertiary treatment of wastewater to remove TDS from treated wastewater. Generally, demineralization is carried out through either resin technology or membrane technology. Resin technology is based on the principle of ion exchange. It is an old technology in which cation and anion resins are used to remove the TDS from wastewater. The fouling of resin, chemical regeneration, and high operation and maintenance costs are the main constraints in the use of this technology in tertiary treatment of wastewater. The latest technology is the membrane-based RO technology for removal of TDS from wastewater. The flow of water through a semipermeable membrane from the more concentrated solution into the diluted solution under influence of pressure greater than the osmotic pressure is called RO.

3.4.3.1 Reverse Osmosis

RO is the technology of choice for salty water treatment. To ensure optimum performance of RO systems, appropriate pretreatment must be selected. All raw waters contain microorganisms: bacteria, algae, fungi, viruses, and micron impurities. The entry of these microorganisms and micron impurities in RO systems causes fouling of membranes. The fouling leads to an increase in the differential pressure from feed to concentrate, finally leading to membrane flux declination and mechanical damage of the membrane.

Fouling prevention is therefore a major objective of the pretreatment process. The potential for fouling is higher with surface water than with well water. The possible measures against fouling – conventional pretreatment and photocatalytic detoxification-based pretreatment – are discussed in this section.

3.4.3.1.1 Reverse Osmosis and Pretreatment

The flow of water through a semipermeable membrane from the more concentrated solution into the diluted solution under the influence of pressure greater than the osmotic pressure is called RO. The process diagram in Figure 3.18 shows a simple way how an RO plant is designed.

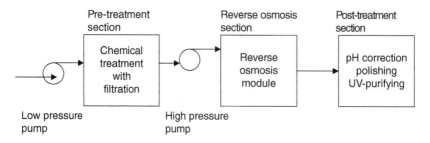

Figure 3.18 Process diagram of reverse osmosis plant.

To increase the efficiency and life of an RO system, effective pretreatment of the feed water is required. Selection of the proper pretreatment will minimize fouling, scaling, and membrane degradation. The result is the optimization of product flow, salt rejection, product recovery, and operating cost. Fouling prevention techniques of RO systems against various foulants are shown in Table 3.3.

Table 3.3 Fouling prevention techniques of RO systems.

Parameters	Desired/optimum limit	If beyond limit?
Turbidity	<1 NTU	Use sand filtration
Silt density index	<3	Use clariflocculation
Hardness	<170 ppm	Use softener/antiscalant system
Iron/heavy metals	<0.3 ppm	Use iron removal filtration
Organics	Nil	Use chlorination/ultrafiltration
O&G	Nil	Use oil adsorption media
Residual chlorine	Nil	Use activated carbon/sodium metabisulfate dosing
Oxidizing materials	Nil	Use photocatalytic detoxification
Color	Colorless	Use activated carbon

3.4.3.1.2 *Conventional Fouling Prevention Technique*

Conventional fouling prevention technique is a popular method to inactivate most pathogenic microorganisms quickly and reduce micron impurities from raw water. It involves chlorination, filtration, and dechlorination. The conventional pretreatment method of RO membranes is shown in Figure 3.19.

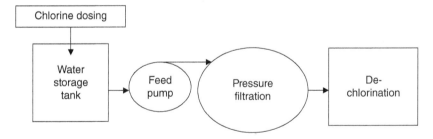

Figure 3.19 Schematic diagram of conventional biological fouling prevention technique.

3.5 Sludge Dewatering

Concerning sludge treatment process selection, the "zero landfill" concept should be applied so that no waste is sent for disposal to landfill or to incineration without energy recovery: therefore waste is either totally eliminated at source or recycled off site. The following considerations should be accounted for regarding the streams that can be produced during the wastewater treatment operations:

- A stream of solid material is removed during the first screening phase.
- O&G streams are separated during the physical and physical/chemical treatments.
- Sludge streams are removed during physical, physical/chemical, and biological processes.

The small stream of coarse material separated by the coarser screen will be collected in big bags or containers and disposed of as a separate waste. In general, the sludge and O&G streams separated from several treatment units will be combined in a common aerated storage tank, dewatered for volume reduction in a filter press unit that can handle both sludge and oily material, and conditioned in a pug mill with chemicals (if needed, according to the site-specific sludge reuse/disposal option).

This recommended sludge treatment scheme has been developed to simplify to the maximum extent the process scheme and operations, reducing the amount of equipment to install, and choosing treatments with an elevated degree of automation (i.e. filter press, pug mill) in order to reduce the operators' workload.

In the case that chemical sludge is produced by the chemical precipitation process (i.e. dosage of $FeCl_3$ or Al_3SO_4 for phosphorus removal), this sludge stream can also be separated from other sludge streams and dewatered separately if the site-specific sludge reuse/disposal criteria require it, as chemical sludge typically can achieve higher dry solid content (up to 75% with filter press).

The simplified sludge treatment scheme described above (see Figure 3.20) should also include anerobic digestion of sludge upstream of the dewatering step if a high amount of biological sludge and/or primary sludge is produced (not including sludge from chemical precipitation processes, as typically not suitable for anaerobic digestion) (see Figure 3.21).

As an optional treatment, the dewatered sludge stream can be further conditioned with alkaline substances for the sludge to be unsuitable for the survival of microorganisms, typically by mixing a proper dosage of lime with dewatered sludge. Stabilized sludge can then be sent to reuse with the following options:

- Fuel in cement plants.
- Raw materials for brick manufacturing.

Whether no stabilization option will be considered, for the dewatered sludge the following recycling alternatives can be evaluated (feasibility to be verified with local regulations):

- **Composting**: Mixing biosolids with bulking agent (wood chips, straw, etc.) with aerobic degradation in a humus-like material with excellent soil conditioning properties.

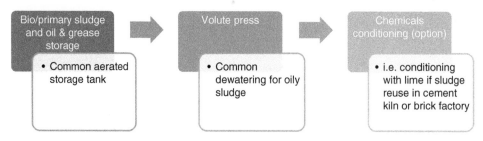

Figure 3.20 Simplified sludge treatment scheme (common treatment for biological/primary sludge and O&G).

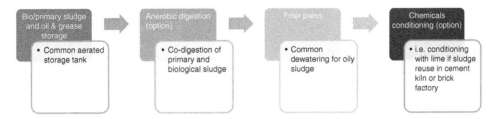

Figure 3.21 Sludge treatment scheme with anerobic digestion (common treatment for biological/primary sludge and O&G).

- **Land application**: Spreading of sludge on soil surface or incorporating or injecting bio-solids into the soil (include application to various types of land including agricultural lands, forests, mine reclamation sites and disturbed lands, parks, and golf courses).
- **Incineration**: Use of biosolids as alternative fuel in incineration plants with energy recovery. This option is feasible especially if additional dewatering can be provided at reasonable costs, i.e. with steam drying beds using available excess heat from the factory.

For some installations it can also be envisaged that the segregation of O&G from the sludge (basically by separating the collecting systems from the gravity separator and from the DAF) could be feasible for:

- Utilization as fuel (thanks to the elevated calorific power).
- Production of animal food or in soap industries (for edible O&G).
- Utilization in the ore-flotation industry, with mineral recovery after acid extraction.

3.5.1 Thickener

The thickening process by simple means of gravity in a static reactor and/or assisted by mechanical scrapers is not recommended for two main reasons:

- If the MBR is the selected biological alternative, MBR waste sludge is typically poorly settleable, so that gravity thickening would not be effective.
- In general, to simplify the wastewater treatment plant layout, sludge and O&G will be collected and treated together, not allowing for efficient gravity separation of supernatant.

Therefore a simple storage tank provided with aeration is recommended (see Figure 3.22).

3.5.2 Sludge Drying Beds

Sludge drying beds (see Figure 3.23) involve low-energy-demanding dewatering and are recommended in the core design after mechanical dewatering units, especially if applied in warm and dry climates with manpower available. The application of sludge drying beds should be evaluated for each site as they:

- Require elevated surfaces and are weather dependent.
- Require manned operations (loading, unloading) and are scarcely automated.
- Do not allow proper control of the dewatering process and can generate odors.

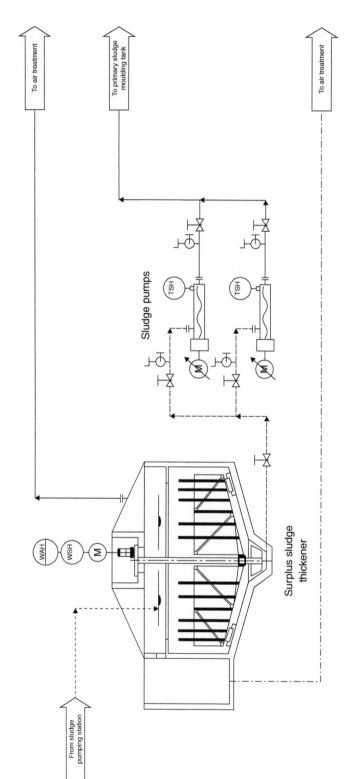

Figure 3.22   Schematic diagram of paddle sludge thickener.

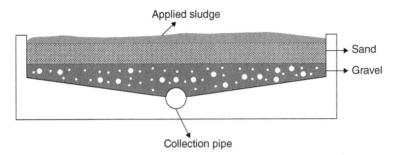

Figure 3.23 Schematic representation of a sludge drying bed.

3.5.3 Anerobic Stabilization

In the core design, anerobic stabilization of sludge is considered an optional treatment only for larger plants where an important amount of biological sludge is produced and where the reduction of sludge to be disposed of offsite is a priority for the specific site. In these cases also primary sludge will be sent for anaerobic digestion if no chemical precipitation is performed.

Anaerobic digestion should be performed in a single-stage high-rate digester, located upstream of the dewatering section and operated at 35 °C. Figure 3.24 shows typical shapes of anaerobic digesters for sludge.

The decision should be based on the following considerations:

- Achievable sludge reduction through anaerobic digestion on the overall amount of sludge produced in the plant (as percentage and amount of TSS reduced).
- Energy balance of the digester, accounting for both the heat required to preheat the feed and the digester heat dispersions.
- Possibility to treat in the digester off-spec food products generated in production and/or food waste from the facility canteen.

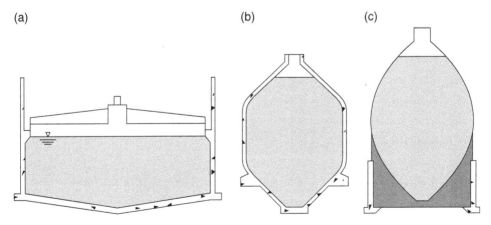

Figure 3.24 Typical shapes of anaerobic digesters for sludge: (a) cylindrical with reinforced concrete construction, (b) conventional, (c) egg-shaped (mainly used in large municipal WWTPs).

Two scenarios have been developed on the basis of the wet sludge temperature:

- Moderate climate, sludge temperature 15 °C.
- Warm climate, sludge temperature 25 °C.

In terms of energy balance in a moderate climate (scenario 1), in some cases the digester will need an external heat source of natural gas to be heated at the operating temperature, as an energy deficit is reported in two cases out of three. As expected, anerobic digestion will be more convenient in warm climates, as in scenario 2 (or if temperature is higher) and the available energy from the biogas is always sufficient to heat the digester.

In terms of sludge reduction for both scenarios, anerobic digestion can provide an important reduction of the sludge to dewater in the order of 44% (assuming that all the sludge produced within the plant is fed to anerobic digestion – no chemical precipitation is performed and no chemical sludge is produced).

In general, sludge reduction through anerobic digestion is not recommended when, due to the use of chemical precipitation and the production of chemical sludge, the amount of sludge fed to the digester is much less than the overall sludge produced in the wastewater treatment plant, resulting in negligible reduction of total sludge volume to dewater and dispose.

3.5.4 Filter Press

Since typically the amount of sludge is not huge in small and medium industries with wastewater treatment plants and the common approach is to combine primary and secondary sludge in dewatering, recommended sludge dewatering treatment is performed with a plate and frame filter press (see Figure 3.25), which has the following advantages compared to other dewatering systems (centrifuge and belt press):

- Allows for higher solid concentration in the sludge cake (up to 70 and 75% dry content respectively for biological and chemical sludge).
- Allows for ease and fully automatic management, including automatic cloth washing.
- Can treat oily sludge.

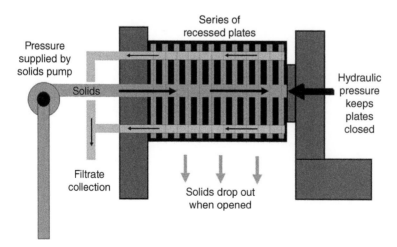

Figure 3.25 Schematic representation of a plate and frame filter press.

If the simplified sludge treatment scheme described in Figure 3.25 is implemented, the filter press can be used to treat both chemical and biological sludge mixed together. If separate treatment of chemical and biological sludge is needed (i.e. if there is a possible different destination selected for the two streams), a filter press unit is needed for each sludge stream as the conditioning dosage and operating cycle will be different according to the sludge characteristics. In cases where separate management of primary and biological sludge is chosen, centrifuge and belt presses could also be considered. Centrifuge could be used for both primary and biological sludge, while belt press can be applied for biological sludge only.

3.5.5 Volute Press

A volute removes water and moisture from sludge on a continuous basis. This multi-disc sludge dewatering press consists of two types of rings: a fixed ring and a moving ring. A screw thrusts the rings together and pressurizes the sludge. Gaps between the rings and screw pitch are designed to gradually get narrower toward the direction of sludge cake outlet and the inner pressure of the discs increases due to the volume compression effect, which thickens and dewaters the sludge. Figure 3.26 shows the working of a typical volute.

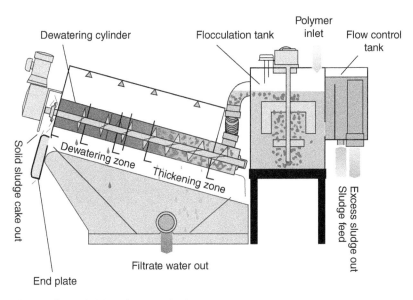

Figure 3.26 Working of a typical volute.

- **Flow control tank**: This feeds a fixed amount of sludge into the flocculation tank. Sludge feed is regulated with the overflow pipe, returning excess volume to the sludge storage tank.
- **Flocculation tank**: Polymer is added with sludge in the flocculation tank for solid/liquid separation. Polymer flocculant and sludge are stirred and mixed, forming flocs suitable for the volute.

- **Cylinder unit**: This mixed sludge passes through a screw cylinder. Filtrate is discharged from the thickening zone and thickened sludge is pushed toward the dewatering zone by screw. In the dewatering zone gaps between rings of the volute screw become smaller toward the end plate.
- **Discharge outlet for dewatered cake**: Further pressure is applied from the outlet side with the end plate, discharging dewatered cake with 20–30% moisture depending on sludge quality.

3.5.5.1   Advantages

- **Easy operation and maintenance**: Intuitively understandable operating system adopted. Monitoring of the operation settings is made very easy. Twenty-four-hour unattended operation is possible with no daily maintenance.
- **Water-saving**: Volute prevents clogging with its unique self-cleaning mechanism, removing the need for huge amounts of water for clogging prevention.
- **Power saving**: The screw which is the main component of the volute rotates very slowly at a rate of 8–10 rpm, so that it consumes very low power and is thus economical. This equipment consume the lowest energy of all dewatering equipment.
- **Low noise/low vibration**: Because volute has no rotating body with high speed, there is no concern about noise and vibration. A comfortable work environment can be secured. During standard operation, noise level is as low as ~65 dB.
- **High resistance to oily sludge**: The self-cleaning mechanism enables volute to be ideal dewatering oily sludge, which easily causes clogging and is difficult to treat with other types of dewatering equipment.
- **Small footprint**: Volute can be installed in places where placement would not be possible with other technologies. This makes it suitable for customers who are considering replacement of existing dewatering equipment.
- **Wide range of application**: Applications including municipal sewage sludge, sludge generated from various industrial effluent treatment plants, primary/chemical sludge, biological sludge, oily sludge, and DAF sludge.

3.6   Zero Liquid Discharge

ZLD is a wastewater management system that ensures that there will be no discharge of industrial wastewater into the environment. It is achieved by treating wastewater through recycling and then recovery and reuse for industrial purpose. Hence ZLD is a closed loop cycle with no discharge.

Water pollution by inappropriate management of industrial wastewater is one of the major environmental problems encountered globally. It is extremely necessary to treat as well as recycle industrial wastewater to reduce water scarcity and decrease the burden on natural resources. Industrial ZLD has received steadily increasing attention across the world due to stringent regulations and water scarcity. To achieve industrial ZLD, innovative and sustainable approaches are required to select the right technology. In Chapter 6 we discuss innovative and sustainable approaches toward industrial ZLD which will help industry to address the issue of minimizing waste as well as help in adoption of cleaner production technologies. Before considering ZLD, understand your treatment goals, economics, and regulatory requirements. For example, concentrating wastewater to a lower volume brine that can be sent for disposal may be more cost effective than producing ZLD solids.

3.6.1 ZLD Technologies

There are several methods available at tertiary stage for the treatment of wastewater to achieve ZLD. Some of the prominent ZLD technologies at tertiary stage for the treatment of wastewater are listed:

- RO.
- Scaleban.
- FO.
- Vacuum distillation.
- Multiple-effect evaporator cum crystallizer.
  ZLD is discussed in Chapter 6 of this book.

References

1 Chaubey, M. and Kaushika, N.D. (2003). Performance analysis of fixed film bioreactor plants for wastewater treatment. *Journal of Industrial Pollution Control* 19 (2): 203–213.

2 Metcalf & Eddy Inc (2014). *Wastewater Engineering: Treatment and Resource Recovery, 5.* New Delhi: Tata McGraw-Hill.

3 CPHEEO. 2013. Manual on Sewerage and Sewage Treatment Systems – 2013. Central Public Health & Environmental Engineering Organization, Ministry of Urban Development, New Delhi.

Further Reading

Côté, P. and Gilliam, T.M., eds. 1989. Environmental aspects of stabilization and solidification of hazardous and radioactive wastes. Papers presented at 4th International Hazardous Waste Symposium on Environmental Aspects of Stabilization/Solidification of Hazardous and Radioactive Wastes, May 3–6, 1987, Atlanta, Georgia. Special technical publication, 1033. American Society of Testing and Materials, Baltimore, MD.

Crittenden, J.C., Rhodes Trussell, R., Hand, D.W. et al. (1985). *M.W.H.'s Water Treatment, Principles and Design*, 3e. New York: Wiley.

Culp, G., Wesner, G., Williams, R., and Hughes, M.V. (1980). *Wastewater Reuse and Recycling Technology*. Park Ridge, NJ: Noyes Data Corporation.

De Renzo, D.J. 1978. Units Operations for Treatment of Hazardous Industrial Wastes. Pollution Technology Review No. 47. Noyes Data Corporation, Park Ridge, NJ.

Eble, K.S. and Feathers, J. 1992. Water reuse within a refinery. Betz Industrial, Technical paper presented at Conference of National Petroleum Refiners Association, March 22–24.

Grady, C.P.L., Daigger, G.T., and Lim, H.C. (1999). *Biological Wastewater Treatment, 2*. New York: Marcel Dekker.

Hall, D.W., Sandrin, J.A., and McBride, R.E. (1990). An overview of solvent extraction technologies. *Environmental Progress* 9 (2): 98–105.

Nalco Chemical Company (1988). *The Nalco Waterbook, 2*. Napperville, IL: Nalco Chemical Company.

Nemerow, N.L. and Dasgupta, A. (1991). *Industrial and Hazardous Waste Treatment*. New York: John Van Nostrand Reinhold.

Perry, R.H., Green, D.W., and Maloney, J.O. (eds.) (1997). *Perry's Chemical Engineer's Handbook*, 7e. New York: McGraw Hill.

Rosain, R.M., Davis, M.W., York, R.J. et al. (1992). *Wastewater Treatment Manual for Coal Gasification–Combined-Cycle Power Plants*. Palo Alto, CA: Electric Power Research Institute.

Von Palmer, S.A.K., Breton, M.A., Nunno, T.J., Sullivan, D.M., and Surprenant, N.F. 1988. Metal Cyanide Containing Wastes – Treatment Technologies. Pollution Technology Review No. 158. Noyes Data Corporation, Park Ridge, NJ.

Zhongxiang, Z. and Yi, Q. (1991). Water saving and wastewater reuse and recycle in China. *Water Science and Technology* 23: 2135–2140.

4

Design Considerations

CHAPTER MENU

Wastewater Treatment Technologies: Design Considerations, First Edition. Mritunjay Chaubey.
© 2021 John Wiley & Sons Ltd. Published 2021 by John Wiley & Sons Ltd.

This chapter provides brief descriptions of design considerations of various wastewater treatment technologies.

4.1 Screening

A screening unit is used to trap large and small floating particles in wastewater (WW). Such units are adopted as standard for wastewater treatment plants (WWTPs). In conjunction with the screening unit, if needed, an oil and grease (O&G) separator may be installed. The O&G separator is recommended for preliminary free O&G removal when no dissolved air flotation (DAF) treatment is installed downstream. The specification reported in this chapter should be considered indicative and not exhaustive since it is intended as typical and a specific description of the supply should be developed for each single case.

The screening unit performs the following functions:

- Screening by means of a coarse screen.
- Removal of free O&G (if needed).

The unit should be provided in one single line but with a sparing philosophy considered to have, for each of the most critical rotating equipment, a $1+1$ philosophy, with an installed spare that can be automatically switched into operation in case of failure of the main one, in order to ensure the continuity of the operation of the WWTP. Each unit should include a screening unit complete with a coarse screen and free O&G separator.

4.1.1 Process Parameters

Table 4.1 summarizes the main process parameters for the screening section.

Table 4.1 Main process parameters for screening unit.

Description	Value/notes
Liquid depth of screen chamber	1.2 m
Length of channel at upstream of screen mechanism	Minimum three times width of screen channel
Free board	0.5 m
Angle of inclination of screen	45° to horizontal
Flow velocity through screen	1.0 m/s
Inlet and outlet gates or isolation valve	Sluice gate or isolation valves to be considered to control flow velocity
Bar screen openings (for coarse screening)	5 mm

4.1.2 Process Description

The raw WW is sent through a drain to the screening unit, which performs a coarse screening and free O&G removal if no DAF treatment is installed downstream. The screening unit includes a first screening treatment performed by means of a fixed coarse mechanical bar screen (openings 5 mm). If the WW contains a high amount of free O&G (especially if sticky and viscous), the fine screening section should not be installed as it can easily get clogged. It is the responsibility of the screens vendor to ensure that the selected screens can be applied to the raw WWs. The coarse material separated by the screening section is collected in a dry box container that allows dewatering, and is then reused or disposed offsite. Screens are cleaned automatically at fixed time intervals by mechanical scraper.

In the cases when no DAF unit is provided within the primary treatments, after the screening phase WW flows by gravity into a chamber provided with baffles, installed in conjunction with the screening unit, where the free O&G fraction is separated by gravity. The free O&G stream is then removed with a skimmer and pumped to the oil/sludge storage tank by means of external horizontal centrifugal pumps. The skimmer consists of a horizontal slotted pipe located before the baffle. The pipe is periodically rotated by a lever in order to submerge the slots under the water level and allow the floated material to be carried into the pipe. A pH analyzer should be provided before the O&G removal section for pH monitoring of the raw WW.

The required volume of the chamber should be calculated on the basis of the assumed hydraulic retention time (HRT) and on the influent flow to be treated:

$$\text{O\&G separator volume } \left(m^3\right) \text{ required} = \text{Influent flow } \left(m^3/h\right) \times \text{HRT } \left(\text{i.e.} 15 \min\right) / 60 \left(\min/h\right)$$

The WW pretreated by the screening unit is collected in a pit and fed to the following equalization unit by means of a centrifugal pump (one duty, one spare) provided with a variable frequency driver (VFD) receiving inputs from a level indicating control (LIC) located in the pit.

The unit will have integral level controls and instrumentation for the proper automation through a programmable logic controller (PLC) system.

4.2 Equalization Unit

The equalization unit will have the following functions:

- Reduction of possible peaks of hydraulics flow in the influent water.
- Reduction of possible load variations in the influent water.

The unit will be provided in one single line but with a sparing philosophy considered to have, for each of the most critical rotating equipment, a 1 + 1 philosophy, with an installed spare that can be automatically switched into operation in case of failure of the main one, in order to ensure the continuity of the operation of the WWTP. The equalization unit will include:

- One equalization tank.
- All the electromechanical and instrumentation and control (I&C) equipment (pumps, compressor, mixers, piping, valves, instruments, etc.) necessary for proper operation.

4.2.1 Process Parameters

Table 4.2 summarizes the main process parameters for the equalization tank.

Table 4.2 Main process parameters for equalization tank.

Description	Value/notes
Equalization tank capacity	Based on 12–24 h HRT
Construction of equalization tank	Two compartments
Free board	0.5–1.0 m
Air mixing flow rate	0.9–1 Nm^3 per m^3 of tank capacity
Types of diffusers used	Coarse or fine bubble
Recirculation pipelines provision	From equalization pump discharge circulation line must be considered
Off-spec tank provision	An additional tank must be considered for off-spec effluent with 12 h HRT

4.2.2 Process Description

The equalization unit will be installed in order to:

- Reduce possible peaks of hydraulics flow.
- Reduce possible peaks in pollutants concentration in the water fed to downstream units.

The WW will come directly from the screening unit, and it will flow to the equalization tank which will be provided with two submersible mixers (one duty, one stand by) to ensure complete homogenization of the contents of the tank and to avoid total suspended solids (TSS) sedimentation. The mixers should be designed to avoid excessive turbulence and to prevent any volatile organic compounds (VOC) stripping if the WW contains any VOC.

The level of the equalization tank will change in order to compensate for the fluctuation in incoming flow and to allow the downstream sections to be operated under flow indicator control (FIC). A level indicating transmitter (LIT) with low and high alarm will be provided for the equalization unit. The equalization tank will be operated with a variable level in the range 30–80% of the total water level.

The equalized water will be transferred in normal operation under flow control mode (FIC set at fixed flow rate acting on VFD of the pump) to the downstream physical/chemical treatment by means of external centrifugal pumps (one duty, one spare).

4.3 Dissolved Air Flotation

The DAF unit will be the preferred treatment to remove TSS and O&G (and eventually phosphorus) to be installed downstream of the equalization unit when O&G concentration in raw WW is higher than 100 mg/l. If surfactants are present in high concentration in the WW to be treated, DAF treatment should be avoided in order to prevent foam formation issues and a clariflocculation system should be preferred.

The DAF unit performs the following operations:

- O&G, fats, and proteins removal (with de-emulsification of emulsified O&G).
- TSS removal (with efficiency to be determined based on the upstream TSS content).
- Chemical removal of phosphorus (when needed).

The unit will be provided in one single line but with a sparing philosophy considered to have, for each of the most critical rotating equipment, a $1+1$ philosophy, with an installed spare that can be automatically switched into operation in case of failure of the main one, in order to ensure the continuity of the operation of the WWTP. The DAF unit should include:

- pH control system.
- Coagulation-flocculation tanks.
- Sedimentation and flotation basin.
- Air saturation unit.
- All the electromechanical and I&C equipment (pumps, compressor, mixers, piping, valves, instruments, etc.) necessary for proper operation.

4.3.1 Process Parameters

Table 4.3 summarizes the main process parameters for the DAF unit.

Table 4.3 Main process parameters for DAF units.

Parameter	Units	Value
HRT in DAF unit	min	50
Hydraulic load in DAF unit	m/h	3–4
COD total removal efficiency	%	30–60
Saturation vessel pressure	bar	3.5–5.5
Percentage of influent flow to be recycled	%	50
Air to solids ratio (A/S)	kg/kg of solids	0.02

4.3.2 Process Description

The WW coming out from the equalization unit is transferred to the DAF unit which performs the following operations:

- De-emulsification of emulsified O&G and removal of free O&G by means of air bubble flotation.
- TSS removal for gravity settling.
- Chemical removal of phosphorus (if needed).

The DAF unit will consist of the following main components:

- pH adjustment tank with rapid mixer for the dosage of lime solution ($Ca(OH)_2$) or hydrochloric acid solution (HCll).
- Coagulation tank with rapid mixer for dosage of ferric chloride ($FeCl_3$) (used both as a coagulant and for phosphorus removal, if phosphorus chemical precipitation is needed).
- Flocculation tank with slow mixer for flocs formation and dosage of:
  - de-emulsifier to transform emulsified O&G into free O&G;
  - polymer to allow proper flocculation.
- Air flotation unit with oil/sludge removal systems.
- Saturation unit (pressure tanks and related compressor) to dissolve air into water.
- Saturation pumps for the recycled water.

Table 4.4 Chemicals application with pH range.

Chemical	Dosage range (mg/l)	pH
Lime ($Ca(OH)_2$)	150–500	9–11
Ferric chloride ($FeCl_3$)	35–150	4–7
Cationic polymers	2–5	Not affected
Anionic polymers	0.25–1	Not affected

To allow for the air to dissolve in water, a quote of the effluent is recycled and held under pressure in a pressure retention tank where air is injected. Air flow is controlled by a pressure control which regulates a valve on the air line. When water leaves the pressure tank and is mixed with the unpressurized untreated stream that enters the DAF tanks, air is released from solution in the form of fine bubbles and rises to the top of the tank, capturing the floating materials.

In the coagulation-flocculation chamber the pH will be adjusted to the optimal operating value to destabilize the emulsions. The choice between the dosage of $Ca(OH)_2$ and HCl solutions will be based on the pH of the incoming stream. The operating pH value and the chemical dosage should be adjusted on the basis of specific jar tests. A pH set-point control is implemented in the DAF tanks to provide the inputs to the relative dosing pumps. As an alternative, the coagulation-flocculation chamber can be replaced with a pipe flocculation configuration with an in-line static mixer installed. A de-emulsifier solution is dosed by means of a dosing pump in order to improve the emulsion destabilization and the phase separation between oil and water.

If phosphorus removal is needed, it will be achieved by chemical precipitation by dosing iron salts ($FeCl_3$). Phosphate is incorporated in TSS and removed with the sludge. $FeCl_3$ will be dosed by means of metering pumps controlled with a loop on the incoming flow rate (the dosage rate will be adjusted on the basis of specific jar tests).

In order to promote the aggregation of larger flocks to be separated in a much more easy way, a polymer solution will be dosed by means of metering pumps controlled based on the incoming flow (the dosage rate will be proportional to the incoming flow and determined with jar-tests). Generally, Table 4.4 can be followed for chemicals application with pH range.

The collected oil will be pumped to the oil/sludge storage tank by means of external horizontal centrifugal pumps (one duty, one spare). The solids will be scraped into the central hopper of the tank and pumped by means of screw pumps (one duty, one spare) to the oil/sludge storage tank. The DAF effluent will flow by gravity to a pit and will be fed to the downstream biological unit by means of a set of external centrifugal pumps (one duty, one spare). The pump will be provided with VFD receiving inputs from a LIC located in the pumping pit.

4.4 Clariflocculator

Clariflocculator reduces suspended solids from effluent with the help of chemicals. The clariflocculator, installed downstream of the equalization unit, is an alternative treatment to the DAF process to remove TSS and O&G (and eventually phosphorus) when O&G concentration in raw WW is lower than 100 mg/l. Moreover, it is considered a reliable alternative when high surfactant concentrations could create foaming issues for the DAF due to the presence of micro air bubbles.

The clariflocculator performs the following operations:

- TSS removal (with efficiency to be determined based on the upstream TSS content).
- O&G removal (obtained by de-emulsification of emulsified O&G).
- Chemical removal of phosphorus (when needed).

The unit will be provided in one single line but with a sparing philosophy considered to have, for each of the most critical rotating equipment, a $1+1$ philosophy, with an installed spare that can be automatically switched into operation in case of failure of the main one, in order to ensure the continuity of the operation of the WWTP.

The clariflocculator should include:

- Upstream coagulation.
- Coagulation chamber.
- Flocculation-sedimentation chambers.
- pH control system.
- All the electromechanical and I&C equipment (pumps, compressor, mixers, piping, valves, instruments, etc.) necessary for proper operation.

4.4.1 Process Parameters

Table 4.5 summarizes the main process parameters for the clariflocculator.

Table 4.5 Main process parameters for clariflocculators.

Parameter	Units	Value
HRT in coagulation chamber (minimum–maximum)	min	5–10
HRT in flocculation chamber (minimum–maximum)	min	10–20
HRT in sedimentation chamber	min	50
Hydraulic load in sedimentation chamber	m/h	1.5

4.4.2 Process Description

The WW coming from the equalization unit is transferred to the clariflocculator which performs the following operations:

- Coagulation (with addition of $FeCl_3$ or $Al(SO_4)_3$), flocculation (with polymer dosage), and gravity settling of TSS.
- De-emulsification and removal of emulsified O&G by means of dosage of specific de-emulsifier.
- Chemical removal of phosphorus (if needed).

Due to possible presence of foam in the WW, a defoaming system can be installed upstream of the clariflocculator in order to reduce the effect. In this case, an antifoaming agent is dosed in the same tank where pH adjustment and coagulation take place.

The clariflocculator will consist of the following main components:

- Coagulation tank with rapid mixer for dosage of:
  - ferric chloride ($FeCl_3$) or alum (used as a coagulant) and for phosphorus removal (if phosphorus chemical precipitation is needed);
  - antifoaming agent to avoid foaming issues;
  - lime solution ($Ca(OH)_2$) or HCl for pH adjustment.
- Flocculation-sedimentation chambers, as part of the clariflocculator unit, where flocculation and subsequent gravity separation take place. In the flocculation chamber, polymer (to allow proper flocculation) and de-emulsifier (to transform emulsified O&G into free O6G) are added.

In the coagulation-flocculation chamber the pH will be adjusted to the optimal operating value to destabilize the emulsions. The choice between the dosage of $Ca(OH)_2$ and HCl solutions will be based on the pH of the incoming stream. The operating pH value and the chemical dosage should be adjusted on the basis of specific jar tests.

A pH set-point control is implemented to provide the inputs to the relative dosing pumps.

A de-emulsifier solution is dosed by means of a dosing pump in order to improve the emulsion destabilization and the phase separation between oil and water.

If phosphorus removal is needed, it will be achieved by chemical precipitation by dosing iron salts ($FeCl_3$). Phosphate is incorporated in TSS and removed with the sludge. $FeCl_3$ will be dosed by means of metering pumps controlled with a loop on the incoming flow rate (the dosage rate will be adjusted on the basis of specific jar tests).

In order to promote the aggregation of larger flocks to be separated in a much more easy way, a polymer solution will be dosed by means of metering pumps controlled based on the incoming flow (the dosage rate will be proportional to the incoming flow and determined with jar tests).

The collected oil will be pumped to the oil/sludge storage tank by means of an external horizontal centrifugal pump (one duty, one spare). The solids will be scraped into the central hopper of the tank and pumped by means of a screw pump (one duty, one spare) to the oil/sludge storage tank. The clariflocculator effluent will flow by gravity to a pit and will be fed to the downstream biological unit by means of an external centrifugal pump (one duty, one spare). The pump will be provided with a VFD receiving inputs from an LIC located in the pumping pit. The unit will have integral level controls and instrumentation for the proper automation through a dedicated PLC system.

4.5 Conventional Activated Sludge

The conventional activated sludge (CAS) unit is one of the selected alternatives for the biological treatment of WW treatment plants. The CAS unit performs two unitary operations:

- Biological reaction, performed in the anoxic and oxidation treatment tanks.
- Solid/liquid separation by settling of biological sludge, achieved in a secondary clarifier.

The unit will be provided in one single line but with a sparing philosophy considered to have, for each of the most critical rotating equipment, a 1 + 1 philosophy, with an installed spare that can be automatically switched into operation in case of failure of the main one, in order to ensure the continuity of the operation of the WWTP. The unit should include:

- The aerated biological reactor provided with an initial anoxic selector.
- The secondary clarifier unit.
- All the electromechanical and I&C equipment (pumps, compressor, mixers, piping, valves, instruments, etc.) necessary for proper operation.

4.5.1 Process Parameters

Table 4.6 summarizes the main process parameters for the biological reactors based on the activated sludge process.

Table 4.6 Main process parameters for biological reactors.

Parameter	Units	Value
TSS concentration in aeration	g TSS/l	3–4
DO concentration in aeration	mg/l	1.5–2.5
F/M ratio in aeration	kg CODtot/kg TSS/day	0.20 (max)
Volumetric load in aeration	kg CODtot/m^3/day	0.8
VSS/TSS ratio		0.8
Sludge age	days	>15
Anoxic selector HRT	min	≥20
Anoxic/aerobic tanks water level	m	5.0–5.5
Biological sludge production (useful to assess WAS flow rate)	kg TSS/day	0.25
Phosphorus for bacteria cells synthesis (in WAS)	g Ptot/kg TSS	0.01
Nitrogen for bacteria cells synthesis (in WAS)	g N/kg TSS	0.12
Mixed liquor suspended solids (MLSS)	ppm	3000–4000
Free board	m	0.6–1

4.5.2 Sizing of Activated Sludge Treatment Systems

In an activated sludge treatment system based on effluent characteristics we select any of following units or combination of all units:

- Activated sludge unit (ASU).
- Denitrification nitrification biotreater (DNB).
- Intermittent nitrification/denitrification biotreater (INDB).

Sizing of ASU, DNB, and INDB are discussed below.

4.5.2.1 Sizing of Activated Sludge Unit

In the ASU, organic pollutants are removed biologically by oxidizing the organic matter into CO_2, H_2O, and new biomass (sludge). Dependent on the ammonia and total Kjeldahl

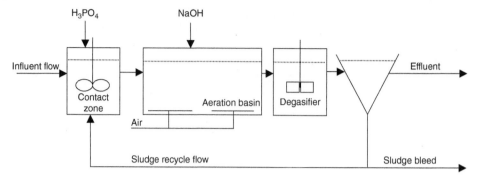

Figure 4.1 Schematic diagram of activated sludge unit.

nitrogen (TKN) demand, oxidation of ammonia to nitrate occurs as well. The ASU consists of the following main process units (see Figure 4.1):

- **Contact zone**: In the contact zone the incoming influent is mixed with recycled sludge. The selector is equipped with a mechanical mixer. The function of the contact zone is to favor conditions for the development of good settling sludge by promoting floc-forming organisms over filamentous organisms. The latter are known to cause a bad settling bulking sludge.
- **Aeration basin**: In the aeration basin organic pollutants (BOD, COD) are biologically oxidized to CO_2, H_2O, and new sludge using O_2. Oxidation of sulfide (to sulfate) and/or ammonia (to nitrate), called nitrification, may occur as well. The aeration basin is continuously aerated and mixed so that O_2 is supplied and the content is kept homogenous. Aeration/mixing is done by aeration equipment (usually bubble aeration). Mechanical mixing is normally not required.
- **Degasifier**: In the degasifier air bubbles attached to the sludge are released and the sludge is allowed to flocculate. A slow paddle mixer supports these processes.
- **Clarifier**: The sludge water mixture enters the clarifier where sludge and water are separated by sedimentation. The settled sludge is recycled; a small amount is discharged. The treated water, virtually free from solids, is discharged or further treated.

4.5.2.1.1 No Nitrification

Contact zone: Design criterion is contact time. Recommended range 5–15 minutes.

$$\text{Volume} = \left[\text{Qinfl} + \text{Qrec} \right] / \left[CT_{SEL} * 60 \right]$$

where

$\quad$ Qinfl, Qrec = Influent or sludge recycle flow in m^3/h
$\quad$ CT_{SEL} = Contact time selector in minutes

Aeration basin: Design criterion is food to microorganism (F/M) ratio. Recommended for extended aeration.

$\quad$ F/M ratio ≤ 0.15 kg BOD/kg MLVSS/day

$$\text{Volume} = \text{BOD} / \left[F / M * \text{MLVSS} \right]$$

where

   BOD = BOD load in kg/day
   MLVSS = Sludge concentration in kg MLVSS/m^3

Degasifier: Design criterion is contact time. Recommended range 10–20 minutes.

$$\text{Volume} = \left[\text{Qinfl} + \text{Qrec} \right] / \left[\text{CT}_{\text{DEG}} * 60 \right]$$

where

   CT_{DEG} = contact time in degasifier in minutes

Clarifier: Design criteria are surface load, sludge concentration, and sludge volume index (SVI).

where

   SL = 450 / [SVI * MLSS]
   A = Qinfl / SL
   D = $\sqrt{[4 * A / (n * \Pi)]}$
   SL = Surface load
   A = Surface
   n = Number of clarifiers

4.5.2.1.2 Nitrification

Contact zone: Design criterion is contact time. Recommended range 5–15 minutes.

$$\text{Volume} = \left[\text{Qinfl} + \text{Qrec} \right] / \left[\text{CT}_{\text{SEL}} * 60 \right]$$

where

   Qinfl, Qrec = Influent or sludge recycle flow in m^3/h
   CTSEL = Contact time selector in minutes

Aeration basin: Design criteria are sludge age and F/M ratio.

Recommended F/M ratio ≤0.15 kg BOD/kg MLVSS/day
Recommended sludge age dependent on temperature.
Sludge age Volume 1 = SP * SRT/ MLSS
F/M ratio Volume 2 = BOD (kg/day) / [F/M * MLVSS (kg/m^3)]
Actual design volume = Max (volume 1, 2)
SP = Sludge production in kg/day
SRT = Sludge age, sludge retention time in days
MLSS = Sludge concentration in kg MLSS/m^3

Degasifier: Design criterion is contact time. Recommended range 10–20 minutes.

$$\text{Volume} = \left[\text{Qinfl} + \text{Qrec} \right] / \left[\text{CT}_{\text{DEG}} * 60 \right]$$

where

   CT_{DEG} = Contact time in degasifier in minutes

Clarifier: Design criteria are surface load, sludge concentration, and SVI.

SL = 450 / [SVI * MLSS]
A = Qinfl / SL
D = $\sqrt{[4 * A / (n * \Pi)]}$
SL = Surface load
A = Surface
n = Number of clarifiers

4.5.2.2 Sizing of Denitrification/Nitrification Biotreater

In the DNB organic pollutants (BOD) are removed biologically by oxidizing the organic matter into CO_2, H_2O, and new biomass (sludge). In addition, nitrogen is removed by nitrification/denitrification. Nitrification and denitrification is carried out in separate reactors. The DNB consist of the following main process units (see Figure 4.2):

- **Contact zone**: In the contact zone the incoming influent is mixed with recycled sludge and recycled mixed liquor. The contact zone is equipped with a mechanical mixer. The function of the selector is as in the ASU.
- **Anoxic basin**: In the anoxic basin nitrate is denitrified to N_2 (denitrification) and BOD is oxidized to CO_2, H_2O, and new sludge. BOD is oxidized with NO_3 instead of O_2. The anoxic tank is equipped with mechanical mixers.
- **Aeration basin**: In the aeration basin BOD not removed in the anoxic tank is oxidized to CO_2, H_2O, and new sludge using O_2. In addition ammonia is oxidized to nitrate (nitrification). For mixing/aeration, aerators (usually bubble aeration) are used. Mechanical mixing is normally not required.
- **Degasifier**: In the degasifier air bubbles attached to the sludge are released and the sludge is allowed to flocculate. A slow paddle mixer supports these processes.
- **Clarifier**: The sludge water mixture enters the clarifier where sludge and water are separated by sedimentation. The settled sludge is recycled; a small amount is discharged. The treated water, virtually free from solids, is discharged or further treated.

Contact zone: Design criterion is contact time. Recommended range 5–15 minutes.

$$\text{Volume} = \left[\text{Qinfl} + \text{Qrec} + Q_{MLR} \right] / \left[CT_{SEL} * 60 \right]$$

where

Qinfl, Qrec = Influent or sludge recycle flow in m^3/h
QMLR = Mixed liquor recycle in m^3/h
CTSEL = Contact time selector in minutes

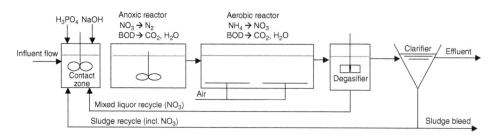

Figure 4.2 Schematic diagram of denitrification/nitrification biotreater unit.

Anoxic basin: Design criteria are denitrification rate and minimal contact time. Recommended minimal contact time 20–60 minutes. Denitrification rate is dependent on temperature.

$$Denitrification\ rate\ Volume1 = N_{DENIT} / \left[DN\text{-}rate * MLVSS \right]$$

where

Contact time Volume 2 = [Qinfl + Qrec + Q_{MLR}] / [CT_{ANOXIC}]
Actual design Vol-anox = Max (volume 1, 2)
N_{DENIT} = Amount of N to be denitrified in kg/day
DN-rate = Denitrification rate in kg N/kg MLVSS/day
MLVSS = Sludge concentration in kg MLVSS/m^3
CT_{ANOXIC} = Contact time in anoxic basin in hours

Aeration basin: Design criterion is aerobic sludge age or total sludge age. Calculated aeration volume is checked with F/M ratio. If F/M ratio is too high the calculated volume is increased to be in compliance with F/M ratio.

Recommended F/M ratio $\leq$0.15 kg BOD/kg MLVSS.d. Recommended sludge ages are dependent on temperature.

4.5.2.2.1 Aerobic Sludge Age

$$Vol-aer = SP * SRT_{AER} / MLSS$$

Check actual F/M ratio = BOD / [(Vol-anox + Vol-aer) * MLVSS] $\leq$0.15where

SP = Sludge production in kg/day
SRT_{AER} = Aerobic sludge age in days
MLSS = Sludge concentration in kg MLSS/m^3
BOD = BOD load in kg/day
MLVSS = Sludge concentration in kg MLVSS/day

4.5.2.2.2 Total Sludge Age

$$Vol-aer + Vol-anox = SP * SRT_{TOT} / MLSS \rightarrow Vol-aer$$
$$= SP * SRT_{TOT} / MLSS - VOL-anox$$

where:

SRT_{TOT} = Total sludge age in days

Degasifier: Design criterion is contact time. Recommended range 10–20 minutes.

$$Volume = \left[Qinfl + Qrec + Q_{MLR} \right] / \left[CT_{DEG} * 60 \right]$$

where

CT_{DEG} = Contact time in degasifier in minutes

Clarifier: design criteria are surface load, sludge concentration, and SVI.

SL = 450 / [SVI * MLSS]
A = Qinfl / SL

$D = \sqrt{[4 * A / (n * \Pi)]}$

SL = Surface load

A = Surface

n = Number of clarifiers

4.5.2.3 Sizing of Intermittent Nitrification/Denitrification Biotreater

In the INDB organic pollutants (BOD) are removed biologically by oxidizing the organic matter into CO_2, H_2O, and new biomass (sludge). In addition nitrogen is removed by nitrification/denitrification. Nitrification and denitrification is carried out in one reactor. The INDB consists of the following main process units (see Figure 4.3):

- **Contact zone**: In the contact zone the incoming influent is mixed with recycled sludge. The selector is equipped with a mechanical mixer. The function of the contact zone is to favor conditions for the development of good settling sludge by promoting floc-forming organisms over filamentous organisms. The latter are known to cause a bad settling bulking sludge.
- **Aeration/anoxic basin**: In this basin BOD is oxidized to CO_2, H_2O, and new sludge using O_2 and/or NO_3. In addition, ammonia is oxidized to nitrate (nitrification) and denitrified to N_2. To allow for nitrification and denitrification in one reactor the system is alternately aerated and not aerated but mixed. For aeration and mixing the basin is equipped with an aerator (usually bubble aeration) and mechanical mixers.
- **Degasifier**: In the degasifier air bubbles attached to the sludge are released and the sludge is allowed to flocculate. A slow paddle mixer supports these processes.
- **Clarifier**: The sludge water mixture enters the clarifier where sludge and water are separated by sedimentation. The settled sludge is recycled; a small amount is discharged. The treated water, virtually free from solids, is discharged or further treated.

Contact zone: design criterion is contact time. Recommended range 5–15 minutes.

$$\text{Volume} = \left[Qinfl + Qrec \right] / \left[CT_{SEL} * 60 \right]$$

where

Qinfl, Qrec = Influent or sludge recycle flow in m^3/h

CTSEL = Contact time selector in minutes

Aeration/anoxic basin: contrary to DNB system, nitrification/denitrification is carried out in one reactor. This reactor is alternately operating as an aerobic and anoxic reactor.

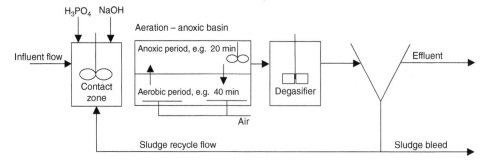

Figure 4.3 Schematic diagram of intermittent nitrification/denitrification biotreater unit.

The effective aerobic volume is then the total volume multiplied by the relative time that the reactor is aerated.

Design criteria are aerobic sludge age or total sludge. Calculated volume is checked with F/M ratio. If F/M ratio is too high the calculated volume is increased to be in compliance with F/M ratio.

Recommended F/M ratio ≤ 0.15 kg BOD/kg MLVSS.d. Recommended sludge ages are dependent on temperature.

4.5.2.3.1 Aerobic Sludge Age

$$\text{Vol} - \text{aer} = \text{SP} * \text{SRT}_{AER} / \text{MLSS}$$

where

SP = Sludge production in kg/day
SRT_{AER} = Aerobic sludge age in days
MLSS = Sludge concentration in kg/m^3
Vol-aer = Required "aerobic volume" in m^3

The total reactor volume is the sum of the required aerobic and the anoxic volume. Both volumes are set by the time (hours/day, minutes/hours) the reactor is aerated (nitrification time) and not aerated but mixed (denitrification time).

The calculation can be iterative. Define denitrification and nitrification time, calculate volumes, check if nitrification and denitrification time are sufficient, and correct if needed.

Check nitrification/denitrification:

$$\text{NiTIME} \geq \text{NNIT} / \left[\text{MLVSS} * \text{VOL} - \text{tot} * \text{NiRATE} \right]$$

$$\text{DeNTIME} \geq \text{NDENIT} / \left[\text{MLVSS} * \text{VOL} - \text{tot} * \text{DeNRATE} \right]$$

where

NNIT = N to nitrify in kg N/day
NiRATE = Nitrification rate in kg N/kg MLVSS.d
NDENIT = N to denitrify in kg N/day
DeNRATE = Denitrification rate in kg N/kg MLVSS.d
MLVSS = Sludge concentration in kg/m^3

Nitrification and denitrification rates are temperature dependent. Check F/M ratio.

$$\text{F/M ratio} = \text{BOD} / \left[\text{Vol} - \text{tot} * \text{MLVSS} \right] \leq 0.15$$

4.5.2.3.2 Total Sludge Age

$$\text{Vol} - \text{tot} = \text{SP} * \text{SRTTOT} / \text{MLSS}$$

$$\text{Vol} - \text{anox} = \text{Vol} - \text{tot} * \text{DeNTIME} / 24$$

$$\text{Vol} - \text{aer} = \text{Vol} - \text{tot} - \text{Vol} - \text{anox}$$

where

SRTTOT = Total sludge age in days

Degasifier: design criterion is contact time. Recommended range 10–20 minutes.

$$\text{Volume} = \left[\text{Qinfl} + \text{Qrec} \right] / \left[\text{CTDEG} * 60 \right]$$

where

CTDEG = Contact time in degasifier in minutes

Clarifier: design criteria are surface load, sludge concentration, and SVI.

$$\text{SL} = 450 / \left[\text{SVI} * \text{MLSS} \right]$$

where

A = Qinfl/SL
D = $\sqrt{[4 * A/(n * \Pi)]}$
SL = Surface load
A = Surface
n = Number of clarifiers

4.5.2.4 Sizing of Sludge Recycling in ASU, DNB, and INDB

$$\text{Qrec} = \text{Qinfl} * \left(\text{MLSS} \right) / \left[\text{MLSS} - \text{recycle} - \text{MLSS} \right]$$

where

MLSS = Sludge concentration in aeration tank in kg/m^3
MLSS-recycle = Sludge concentration sludge recycle = Min (1200/SVI,10) in kg/m^3
Qrec = Quantity of recycle flows in m^3/h
Qinfl = Quantity of influent flows in m^3/h

4.5.2.5 Sizing of Aeration Requirement in ASU, DNB, and INDB
Aeration is needed to supply O_2 for the biological process.

$$\text{ASU Actual Oxygen demand} \left(\text{AOD} \right) = \text{oxygen BOD removal} + \text{sludge respiration} \\ + \text{oxygen nitrification}$$

where

BOD removal = BOD load × BOD factor (BOD factor = 0.6–0.8)
Sludge respiration = Aeration volume × MLVSS × factor (factor = 0.1–0.15)
Nitrification = N to nitrify × 4.57

$$\text{DNB AOD} = \text{oxygen BOD removal} + \text{sludge respiration} + \text{oxygen nitrification} \\ + \text{oxygen denitrification}$$

where

BOD removal = BOD load × BOD factor (BOD factor = 0.6–0.8)
Sludge respiration = Aeration volume × MLVSS × factor (factor = 0.1–0.15)
Nitrification = N to nitrify × 4.57

Denitrification = N to denitrify × (−2.86)

INDB AOD = oxygen BOD removal + sludge respiration + oxygen nitrification
+oxygen denitrification

where

BOD removal = BOD load × BOD factor (BOD factor = 0.6–0.8)
Sludge respiration = Aeration volume × MLVSS × factor (factor = 0.1–0.15)
Nitrification = N to nitrify × 4.57
Denitrification = N to denitrify × (−2.86)

4.5.3 Process Description

In CAS treatments, COD/BOD removal is accomplished by biomass which grows in a flocculent state suspended in the liquid phase in the bioreactor. The WW inlet is sent to the biological tank in the initial anoxic selector. The anoxic selector is a first step in the hydrolysis of organic matter, equipped with submerged mixers. The mixed liquor flows by gravity from the anoxic selector into the oxidation section through submerged openings.

In the following oxidation tank, in order to supply oxygen for biological activity, air is provided with a set of air lobe blowers (one duty, one standby) and by means of fine bubble air diffusers installed at the bottom of the tanks that maximize the oxygen transfer into the mixed liquor.

The amount of air to be provided will be based on a set-point control set on the dissolved oxygen (DO) meters in the oxidation tank (DO level at 1.5–2.5 mg/l). The DO meter controls the VFD installed on the air blowers. To size the blowers for each specific WWTP, at first the peak oxygen requirement for biological reaction should be estimated.

The daily peak oxygen requirement for biological reaction should be calculated as follows:

$$AOR_{BIO} = a' \times BOD_{5REM} + b' \times MLVSS$$

where

AOR_{BIO} = Oxygen requirement for biological process in kgO_2/day
a' = Coefficient of oxygen demand for synthesis (assumed as 0.5)
BOD_{5REM} = kg BOD_5/day removed in the oxidation tank; if the added COD is easily biodegradable, it can be conservatively assumed that $COD = BOD_5$
b' = Coefficient of endogenous demand (assumed as 0.24 for peak conditions at 30 °C)
MLVSS = Total mixed liquor volatile suspended solids under aeration (kgVSS).
The amount of air to be provided by the blowers in standard conditions can be calculated from the peak oxygen requirement with the following equation:

$$Q_{AIR} = c' \times AOR_{BIO} \times Vr \times \eta_{AER}$$

where

c' = Coefficient for conversion to standard conditions at 20 °C and 1 atm (assumed as 1.6)
Vr = Ratio oxygen : air (assumed as 0.3 kgO_2/Nm^3_{AIR})
η_{AER} = Efficiency of the aeration system (assumed as 30%).

The recycle activated sludge (RAS) ratio from secondary clarifier should be ≥100% so the concentration in the biological reactor is around half of the concentration of the RAS flow (ratio between the flow rate of the recycled sludge and the flow rate of the incoming WW to bioreactor). The mixed liquor TSS concentration in the biological tank should be maintained around at 4 gTSS/l.

In order to guarantee conditions suitable for biomass growth and biodegradation, chemicals will be dosed in the biological reactor if necessary (to be assessed on the basis of site-specific data). Nitrogen and phosphorus will be dosed in the anoxic selector as nutrients for biomass growth (supplied with commercial urea and phosphoric acid solutions).

The pH in the aeration tank will be controlled through NaOH or HCl (or H_2SO_4) solution dosage with a pH control loop based on set-point value at neutral pH in the range 7–7.5. The pH meter (located in the final part of the aeration tank, close to the outfall to the clarifier) is acting on an on–off valve on NaOH and HCl (or H_2SO_4) loop (if installed at the facility) or on the VFD of the NaOH and HCl (or H_2SO_4) metering pump.

4.5.3.1 Secondary Clarifier

The mixed liquor will flow by gravity to the secondary clarifier. Depending on the clarifier configuration, the mixed liquor enters the clarifier and follows a radial flow or a plug-flow pattern. Solid/liquid separation takes place by settling of biological sludge from the liquid stream. Settled sludge is conveyed throughout a bottom scraper mechanism, to a part of the tank connected to the suction side of the sludge pumps.

Part of the solids collected in the secondary clarifier is pumped back to the anoxic selector (RAS solids) by means of a set of centrifugal pumps (one duty, one standby) operated at a fixed flow rate adjustable by operators, so as to keep the desired suspended solids concentration in the aerobic reactor. The excess sludge (WAS solids) is pumped to the sludge/oil storage tank by means of a set of screw pumps (one duty, one standby) operated at a fixed flow rate adjustable by operators. As an alternative, it is possible to operate with only one pump set for RAS and WAS streams. In this case, WAS pumps are replaced with a flow control valve and a flow recorder/totalizer. The feasibility of this configuration should be investigated based on specific hydraulic conditions. Table 4.7 summarizes the main parameters for secondary clarifiers.

Table 4.7 Characteristics of a typical secondary clarifier.

Parameter	Units	Value
Hydraulic load	m/h	0.5
TSS outlet concentration	mg/l	15
Water level (min)	m	3

4.6 Moving Bed Biofilm Reactor

MBBR is a combination of suspended growth and attached growth process. It uses the whole tank volume for biomass growth, with simple floating media, which are carriers for attached growth of biofilms. Biofilm carrier movement is caused by the agitation of air

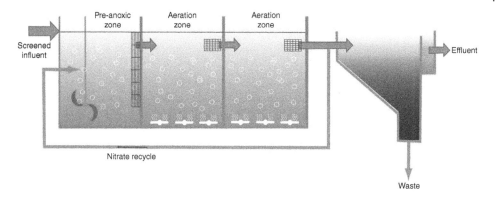

Figure 4.4 MBBR technology process flow diagram.

bubbles. This compact treatment system is effective in removal of organic matter as well as nitrogen and phosphorus, while facilitating effective solids separation. An MBBR technology process flow diagram is shown in Figure 4.4.

4.6.1 Working Principles of MBBR Technology

In the MBBR reactor COD/BOD degradation is performed by attached biomass that grows on the surface of plastic carrier elements. Routine sloughing of biomass from the plastic media caused by the mixing process is not returned to the reactor in MBBR systems but it is all wasted at the same rate at which it leaves the reactor. The WW inlet is sent to the biological tanks in the initial anoxic selector.

The anoxic selector is a first step in the hydrolysis of organic matter, equipped with submerged mixers. The WW flows by gravity from the anoxic selector into the oxidation section through submerged openings. In the following oxidation tank in order to supply oxygen for biological activity, air will be provided with a set of air lobe blowers (one duty, one standby) and by means of coarse bubble air diffusers installed at the bottom of the tanks that maximize the oxygen transfer into the mixed liquor.

The amount of air to be provided will be based on a set-point control set on the DO meters in the oxidation tank (DO level at 2.5–3.5 mg/l). The DO meter controls the VFD installed on the air blowers. To size the blowers for each specific WWTP, at first the peak oxygen requirement for biological reaction should be estimated. The daily peak oxygen requirement for biological reaction should be calculated as follows:

$$AOR_{BIO} = a' \times BOD_{REM} + b' \times MLVSS$$

where

AOR_{BIO} = Oxygen requirement for biological process in kgO_2/day
a' = Coefficient of oxygen demand for synthesis (assumed as 0.5)
BOD_{REM} = kg BOD/day removed in the oxidation tank
b' = Coefficient of endogenous demand (assumed as 0.24 for peak conditions at 30 °C)
MLVSS = Total mixed liquor volatile suspended solids under aeration in kgVSS.

The amount of air to be provided by the blowers in standard conditions can be calculated from the peak oxygen requirement with the following equation:

$$Q_{AIR} = c' \times AOR_{BIO} \times Vr \times \eta_{AER}$$

where

c' = Coefficient for conversion to standard conditions at $20\,°C$ and $1\,atm$ (assumed as 1.6)

Vr = Ratio oxygen : air (assumed as 0.3 $kgO_2/Nm^3{}_{AIR}$)

η_{AER} = Efficiency of the aeration system (assumed as 30%).

In order to guarantee conditions suitable for biomass growth and biodegradation, chemicals will be dosed in the biological reactor if necessary (to be assessed on the basis of site-specific data). Nitrogen and phosphorus will be dosed in the anoxic selector as nutrients for biomass growth (supplied with commercial urea and phosphoric acid solutions).

The pH in the aeration tank will be controlled through NaOH or HCl (or H_2SO_4) solution dosage with a pH control loop based on set-point value at neutral pH in the range 7–7.5. The pH meter, located in the final part of the aeration zone, is acting on an on–off valve on NaOH and HCl (or H_2SO_4) loop (if installed at the facility) or on the VFD of the NaOH and HCl (or H_2SO_4) metering pump.

4.6.2 Process Parameters

MBBR carrier media fill, specific area of carrier media material, dissolved oxygen (DO) concentration in MBBR tank, volumetric load in MBBR tank, MBBR tank water level, phosphorus for bacteria cells synthesis (in WAS), nitrogen for bacteria cells synthesis (in WAS), and temperature are the main process parameters for MBBR technology. Table 4.8 summarizes.

Table 4.8 Main process parameters for MBBR technology.

Parameter	Units	Value
MBBR carrier media fill	%	30–50
Specific area of MBBR carrier media material	m^2/m^3	600–800
DO concentration in MBBR tank	mg/l	2.5–3.5
Volumetric load in MBBR tank	kg CODtot/m^3/day	0.8
MBBR tank water level	m	5–6
Temperature inside MBBR tank	°C	20–25
Phosphorus for bacteria cells synthesis (in WAS)	g Ptot/kg TSS	0.01
Nitrogen for bacteria cells synthesis (in WAS)	g N/kg TSS	0.12
Free board	m	0.6–1

4.6.2.1 Carrier Media

MBBR carrier media (see Figure 4.5) play very a important role in the success of MBBR technology. Critical to the success of any biofilm process is to maintain a high proportion

Figure 4.5 MBBR carrier media – high-density polyethylene.

of active biomass in the reactor. When the biomass concentration on MBBR carriers is presented in terms of an equivalent suspended solids concentration, values typically are approx. 1000–5000 mg/l suspended solids. Yet, when performance is assessed on a volumetric basis, results show that removal rates can be much higher than those compared with suspended-growth systems. This added volumetric efficiency with MBBRs can be attributed to the following:

- High overall biomass activity resulting from effective control of biofilm thickness on the carrier resulting from the shear imparted on the carriers by the mixing energy.
- Ability to retain highly specialized biomass specific to the conditions within each reactor.
- Acceptable diffusion rates resulting from the turbulent conditions in the reactor.
  The net effective biofilm area is a key design parameter with moving-bed reactors, and loading and reaction rates can be expressed as a function of the surface area offered by the carriers. Hence, carrier surface area is convenient and commonly used for expressing performance and loadings of MBBRs often presented as surface area removal rate (SARR) and surface area loading rate (SARL), respectively. The substrate removal rate in MBBRs is zero-order-dependent when bulk substrate concentrations are high and first-order-dependent when bulk substrate concentration is low. Under controlled conditions, removal rate (SARR) as a function of loading (SARL) can be expressed by the following surface reaction rate expression:

$$r = r_{max} * \left[L / \left(K + L \right) \right]$$

where

r = Removal rate in g/m^2/day
r_{max} = Maximum removal rate in g/m^2/day
L = Loading rate in g/m^2/day
K = Half-saturation constant.

4.6.3 Pre-denitrification MBBRs

Pre-denitrification is required where nitrogen removal is required. The influent to the MBBR should have a favorable ratio of easily biodegradable COD and ammonia (C : N) to make efficient use of the anoxic reactor volume. The dissolved oxygen returned in the recycle flow can have a significant influence on performance with MBBR systems, because elevated bulk dissolved oxygen levels can be required for the nitrification stage of the MBBR process. This can place a practical upper limit on the most effective recirculation ratio ($Q_{rcy} : Q_{inf}$), whereby a further increase in the recirculation rate ends up reducing the overall denitrification efficiency. Where influent WW characteristics are suitable for pre-denitrification, nitrogen removal performance typically can range between 50 and 70%, with a $Q_{rcy} : Q_{inf}$ ratio of 1:1 to 3:1.

4.7 Membrane Bioreactor

An MBR unit is one of the selected alternatives for the biological treatment of WW in treatment plants. The MBR unit performs two unitary operations:

- Biological reaction, performed in the anoxic and oxidation treatment tanks.
- Solid/liquid separation achieved through UF membranes submerged in an external aerated reactor that will perform a further oxidation process or through side-stream vertical membranes with air-lift system.

The unit will be provided in one single line but with a sparing philosophy considered to have, for each of the most critical rotating equipment, a 1 + 1 philosophy, with an installed spare that can be automatically switched into operation in case of failure of the main one, in order to ensure the continuity of the operation of the WWTP. The unit should include:

- The aerated bioreactor provided with an initial anaerobic selector.
- The MBR membranes tank.
- The MBR back pulse tank.
- The chemical storage and dosing systems for membranes cleaning.
- All the electromechanical and I&C equipment (pumps, compressor, piping, valves, mixers, instruments, etc.) necessary for proper operation.

4.7.1 Process Parameters

Table 4.9 summarizes the main process parameters for biological reactors.

Table 4.9 Main process parameters for biological reactors.

Parameter	Units	Value
TSS concentration in aeration (for submerged membranes)	g TSS/l	7
TSS concentration in aeration (for external side-stream membranes, air-lift system)	g TSS/l	10
DO concentration in aeration	mg/l	1.5–2.5
F/M ratio in aeration	kg CODtot/kg MLSS/day	0.20 (max)
Volumetric load in aeration	kg CODtot/m³/day	1.4
VSS/TSS ratio		0.8
Sludge age	d	>15

Table 4.9 (Continued)

Parameter	Units	Value
Anoxic selector HRT	min	≥ 20
Anoxic/aerobic tanks water level	m	5.0–5.5
Biological sludge production (useful to assess WAS flow rate)	kg TSS/kg $COD_{oxidized}$	0.25
Phosphorus for bacteria cells synthesis (in WAS)	g Ptot/kg TSS	0.01
Nitrogen for bacteria cells synthesis (in WAS)	g N/kg TSS	0.12
Membrane design-specific flux (for submerged membranes)	$l/m^2/h$	15
Membrane design-specific flux (for external side-stream membranes, air-lift system)	$l/m^2/h$	40

4.7.2 Process Description

4.7.2.1 Biological Reactor

For MBR treatments, COD/BOD removal is accomplished by biomass which grows in a flocculent state suspended in the liquid phase in the bioreactor. The WW inlet is sent to the biological tanks in the initial anoxic selector. The anoxic selector is a first step in the hydrolysis of organic matter, equipped with submerged mixers. The mixed liquor flows by gravity from the anoxic selector into the oxidation section through submerged openings.

In the following oxidation tank, in order to supply oxygen for biological activity, air will be provided with a set of air lobe blowers (one duty, one standby) and by means of fine bubble air diffusers installed at the bottom of the tanks that maximize the oxygen transfer into the mixed liquor.

The amount of air to be provided will be based on a set-point control set on the DO meters in the oxidation tank (DO level at 1.5–2.5 mg/l). The DO meter controls the VFD installed on the air blowers. To size the blowers for each specific WWTP, at first the peak oxygen requirement for biological reaction should be estimated. The daily peak oxygen requirement for biological reaction should be calculated as follows:

$$AOR_{BIO} = a' \times BOD5_{REM} + b' \times MLVSS$$

where

AOR_{BIO} = Oxygen requirement for biological process in kgO_2/day
a' = Coefficient of oxygen demand for synthesis (assumed as 0.5)
$BOD5_{REM}$ = kg BOD_5/day removed in the oxidation tank; if the added COD is easily biodegradable, it can been conservatively assumed that COD = BOD_5
b' = Coefficient of endogenous demand (assumed as 0.24 for peak conditions at 30 °C)
MLVSS = total mixed liquor volatile suspended solids under aeration (kgVSS).

The amount of air to be provided by the blowers in standard conditions can be calculated from the peak oxygen requirement with the following equation:

$$Q_{AIR} = c' \times AOR_{BIO} \times Vr \times \mu_{AER}$$

where

$c' =$ Coefficient for conversion to standard conditions at $20\,°C$ and $1\,atm$ (assumed as 1.6)

$Vr =$ Ratio oxygen : air (assumed as $0.3\ kgO_2/Nm^3_{AIR}$)

$\eta_{AER} =$ Efficiency of the aeration system (assumed as 30%).

The mixed liquor recycle ratio from the membrane tank (if submerged membranes are considered) should be $\geq100\%$ so the concentration in the biological reactor is around half of the concentration of the sludge in the membrane tank (ratio between the flow rate of the recycled sludge and the flow rate of the incoming WW to bioreactor). The mixed liquor TSS concentration in the membrane tank should not exceed $11.5\ g$ TSS/l.

In order to guarantee conditions suitable for biomass growth and biodegradation, chemicals will be dosed in the biological reactor if necessary (to be assessed on the basis of site-specific data). Nitrogen and phosphorus will be dosed in the anoxic selector as nutrients for biomass growth (supplied with commercial urea and phosphoric acid solutions).

The pH in the aeration tank will be controlled through NaOH or HCl (or H_2SO_4) solution dosage with a pH control loop based on set-point value at neutral pH in the range 7–7.5. The pH meter is acting on an on–off valve on NaOH and HCl (or H_2SO_4) loop (if installed at the facility) or on the VFD of the NaOH and HCl (or H_2SO_4) metering pump.

The mixed liquor is then sent to the following solid/liquid separation unit, which may be performed by means of two alternative systems:

- Submerged UF membranes in a separate aeration tank; or
- Side-stream vertical membranes with air-lift system.

The unit will have integral level controls and instrumentation for the proper automation through a dedicated PLC system.

4.7.2.2 Submerged Ultrafiltration Membrane Bioreactors

The mixed liquor will be pumped to the UF membrane bioreactor tanks by means of a set of centrifugal pumps (one duty, one standby) with the suction side connected to the lower part of the bioreactor so as to segregate eventual free oil in the bioreactor tank and avoid the fouling of membranes. Each pump will have a dedicated VFD that will receive the signal from the level meter installed in the oxidation tank, with level control set-point.

The flow will be fed to the membranes by means of a common feed channel provided with a number of equal length weirs equal to the number of tanks (two or three tanks should be provided depending on the unit capacity): each tank will contain one membrane filtering cassette. The type and size of one membrane filtering cassette refers to a typical product commonly available and highly reliable in the international market. The membrane filtering cassette of hollow fiber type will be immersed in the respective tank and performs an outside-to-inside filtration.

Permeate will be extracted by means of the vacuum generated by the external centrifugal suction pumps (one duty per each line and one common standby) and delivered to the following treatment unit. Each permeate pump will be provided with a VFD receiving inputs from a level meter on the membrane tanks and from a flow meter on the delivery side of the membrane feed pumps in order to maintain the flow within a specified range. Characteristics of a typical reference UF membrane bioreactor tank are shown in Table 4.10.

Table 4.10 Characteristics of a typical reference ultrafiltration membrane bioreactor tank.

	Parameter	Units	Value
Membrane tank	Length	m	2.8
	Width	m	2.4
	Total height	m	4.5
	Volume of each tank (min)	m^3	20
	Water depth	m	3–3.5
Membrane cassette	Filtration surface	m^2	1500
	Design-specific flux with all installed trains working	$l/m^2/h$	15
	Maximum specific flux with one train in maintenance	$l/m^2/h$	25

The number of typical standard cassettes should be calculated on the basis of the following parameters:

- The influent flow to be treated (which is variable for each specific plant).
- The filtration surface per each cassette, assumed equal to 1500 m²/cassette.
- The number of modules contained in each cassette, equal to 48 modules/cassette.
- The design-specific flux admissible by the membranes, which should be set equal to 15 l/m²/h in normal conditions, and should generally not exceed 25 l/m²/h with one cassette out of operation due to the cleaning process.

Therefore the calculation should be performed as follows:

- Overall surface (m²) required = Influent flow (m³/h) / Specific flux (15 l/m²/h) x 1000 (in normal operation).
- Number of cassettes required (n cassettes) = Overall surface (m²) / 1500 (m²/cassette) + 1.

The presence of the gates at each tank inlet allows the exclusion of one or more of the tanks when required by the cleaning process or by maintenance purposes; therefore a spare cassette has been included in the above calculation. The cleaning process should be enabled by a closed gate switch. Table 4.11 shows the operational parameters with two cassettes and one cassette.

Each MBR tank is provided with an overflow weir to allow the recycle by gravity of the concentrated mixed liquor back to the biological reactor. The WAS is extracted from the membrane tanks and sent by WAS extraction pumps (one duty, one standby) to the sludge treatment section. Extraction is controlled by means of a flow meter in order to extract daily the correct amount of excess sludge.

The membrane tanks will be aerated to prevent solids settling and to help dislodge solids that may remain attached to the surface of the membranes; also the aeration performs a further oxidation process. Aeration will be accomplished by a dedicated lobe blower (one duty, one standby), included in the unit, which will feed alternatively the membrane tanks.

Table 4.11 Operational parameters with two cassettes and one cassette (1500 m² filtering surface) installed for different flow rates.

Flow rate to membranes		Theoretical membrane surface required @ 15 L/m²/h	Operative specific flux during normal operation (2 cassettes)	Operative specific flux during maintenance (1 cassette in operation, should be <25)
m³/day	m³/h	m²	L/m²/h	L/m²/h
240	10	666.7	3.3	6.7
360	15	1000.0	5.0	10.0
480	20	1333.3	6.7	13.3
600	25	1666.7	8.3	16.6
720	30	2000.0	10.0	20.0
840	35	2333.3	11.7	23.3
960	40	2666.7	13.3	26.7

4.7.2.2.1 Membrane Cleaning

Membranes need to be cleaned on a regular basis. The membrane cleaning procedure will be completely automatic and controlled by the PLC and consist of the following:

- Membrane backwashing with permeate.
- Periodic maintenance chemical cleaning.
- Recovery cleaning.

The cleaning procedure will be fully automated and different in the details depending on the selected contractor. The system will allow in general:

- **Membrane backwash using permeate**: This is achieved by operation of automatic open/close valves and back-pulse pump (one duty, one standby).
- **Periodic maintenance cleaning**: During normal operation, periodic chemical maintenance cleaning will be required. The anticipated frequency is one chemical cleaning cycle per week per tank. The membrane cleaning is accomplished by using sodium hypochlorite solution (200 mg/l) followed by citric acid solution (2000 mg/l). These chemicals are injected on the discharge side of the back-pulse pump during the backwash phase. During the cleaning operation, the respective membrane tank will be offline. This procedure will have duration of approx. 1 hour and 20 minutes.
- **Recovery cleaning**: During recovery cleaning, the membrane tank will be taken out of service, drained, and then filled with permeate and sodium hypochlorite solution to achieve about 1000 mg/l sodium hypochlorite concentration. The tank will be drained again and filled with permeate and citric acid to obtain about 2000 mg/l citric acid solution. The membrane tank will be drained, filled with WW, and then put back online. The recovery cleaning is not anticipated more than two times per year per membrane system.

The chemicals to be used for membrane cleaning are as follows:

- Sodium hypochlorite is used as a disinfectant and oxidant in order to eliminate the microorganisms from the surface of membranes and to eliminate the organic compounds which could foul the membranes.

- Citric acid is used to remove the metallic hydroxides that may have formed on the membrane surface during the sodium hypochlorite wash.

The scope of supply for the vendor should include:

- permeate backwash pumping system (one duty, one standby).
- pumping system for sodium hypochlorite (one duty, one standby) and for citric acid (one duty, one standby);

and may include:

- chemical storage system for both sodium hypochlorite and citric acid.

Detailed procedures for recovery cleaning of the membranes, chemicals, concentrations, and dosages will be specified by the selected membrane vendor. In general, the cleaning sequence controlled by the PLC includes:

- Stop of the permeation in the membrane tank by stopping permeate pumps.
- Backwash using permeate only.
- Transfer of the mixed liquor to the biological tanks by means of the overflow channel.
- Backwash with the permeate and cleaning solution for cleaning.
- Placing the membrane tank in service again.

The unit will have integral level controls and instrumentation for proper automation through a dedicated PLC system.

4.7.2.3 External Side-Stream Ultrafiltration Membranes with Air-Lift System

The mixed liquor will be pumped to the membranes by means of a set of centrifugal pumps (one duty, one standby), with the suction side connected to the lower part of the bioreactor so as to segregate eventual free oil in the bioreactor tank and avoid fouling of membranes. Each pump will have a dedicated VFD that will receive the signal from the level meter installed in the oxidation tank, with level control set-point.

With external, air-lift system configuration, the membranes are placed vertically outside the bioreactor and are not submerged in a specific tank. The membranes are continuously aerated at the bottom of the modules by means of a dedicated blower (one duty, one standby), included in the unit. The aeration will create turbulence that will provide high flux rates and very low transmembrane pressures. Air mixes with the influent mixed liquor and raises the flow through the inner side of the tubular membranes. Therefore air injection is the driving force for sludge recirculation to the bioreactor, so that the feed pump will be used only to overcome hydraulic losses and will be a low pressure type. This will result in lower energy consumption than for submerged MBR systems.

External side-stream membranes are provided in compact modules equipped with tubular type membranes, where filtration is performed with inside-to-outside configuration. Table 4.12 shows typical reference external side-stream UF membranes.

The WAS is extracted from the recirculation loop and sent by WAS extraction pumps (one duty, one standby) to the sludge treatment section. Extraction is controlled by means of a flow meter in order to extract daily the correct amount of excess sludge.

Table 4.12 Ultrafiltration membranes specification.

Parameter	Units	Value
Number of modules per unit installed		6
Membrane area for each module	m^2	55
Total membrane area	m^2	330
Module diameter	mm	0.8

Generally, the advantages of the external air-lift system over submerged configuration are:

- Lower energy consumption;
- Ease in maintenance and cleaning procedures;
- Prevention of fouling on the membrane surface due to air injection.

4.7.2.3.1 Membrane Cleaning

Membranes need to be cleaned on a regular basis. The membrane cleaning procedure will be completely automatic and controlled by the PLC and will consist of a combination of forward flushing and periodic backflushing. In general, the cleaning sequence controlled by the PLC includes:

- Stop of the permeation.
- Backwash using permeate only.
- Transfer of the mixed liquor to the biological tanks.
- Backwash with the permeate and cleaning solution for cleaning.
- Placing the membrane tank in service again.

A permeate storage tank is needed in order to store permeate for backwash operations, performed by means of a backwash pump (one duty, one standby). Periodic chemical maintenance cleaning is required. Therefore chemicals (sodium hypochlorite and hydrochloric acid) are injected on the discharge side of the back-pulse pump during the backwash phase.

The unit will have integral instrumentation for proper automation through a dedicated PLC system.

4.8 Chlorination Unit

A chlorination unit is adopted as a standard unit in WWTPs. The chlorination basin is aimed at removing residual biological activity in order to prevent biofouling in downstream additional tertiary treatments and for the oxidation of residual organics and inorganics by means of sodium hypochlorite dosage.

The unit will be provided in one single line but with a sparing philosophy considered to have, for each of the most critical rotating equipment, a $1+1$ philosophy, with an installed spare that can be automatically switched into operation in case of failure of the main one, in order to ensure the continuity of the operation of the WWTP.

The chlorination basin will include:

- One chlorination tank.
- All the electromechanical and I&C equipment (metering pumps, mixers, piping, valves, instruments, etc.) necessary for proper operation.

4.8.1 Process Parameters

Table 4.13 summarizes the main process parameters for the chlorination tank.

Table 4.13 Main process parameter for chlorination basin.

Parameter	Unit	Value
HRT in chlorination basin	min	30

4.8.2 Process Description

Disinfection by means of sodium hypochlorite is installed in order to:

- remove residual biological activity in order to prevent biofouling in downstream additional tertiary treatments; and
- oxidize residual organic and inorganic compounds.

The process water coming from the biological treatment (or from the granular activated carbon [GAC] unit, if included upstream) will be fed to the chlorination basin where it gets mixed with sodium hypochlorite solution dosed by a metering pump (one duty, one standby). In order to maximize the contact, the basin is provided with one vertical mixer (one duty, one in stock) to ensure complete homogenization.

Sodium hypochlorite dosage will be adjusted to achieve a residual free chlorine concentration that does not affect the downstream treatments, as UF membranes (concentrations to be verified with UF vendors). A chlorine meter based on a set-point control value should be installed in order to monitor residual free chlorine concentration in treated water.

Liquid sodium hypochlorite will be stored at the site in corrosion-resistant vessels, protected from direct sunlight and excessive heat. The amount of chemical stored must be carefully evaluated depending on its usage, because NaOCl progressively loses its strength.

Design of NaOCl storing facility will also account for generation and diffusion of chlorine fumes.

4.9 Pressure Sand Filter

Pressure sand filter (PSF) polishing treatment is a selected tertiary treatment of WW treatment facilities. The PSF unit performs further removal of residual TSS from WW by filtration on sand beds. The PSF should be applied downstream of the biological treatment unit only if the chosen technology for biological treatment is a CAS or MBBR system. No PSF treatment is needed if the chosen biological option is an MBR system.

A sparing philosophy will be adopted considered to have, for each of the most critical rotating equipment, a $1+1$ philosophy, with an installed spare that can be automatically switched into operation in case of failure of the main one, in order to ensure the continuity of the operation of the WWTP.

The filtration on the sand media will be performed into a number of fixed-bed columns, organized to achieve an in-series contact process; the number of filters will depend on the plant size and should in any case include two sand filters in parallel (as minimum) in order to guarantee at least one operating filter during maintenance activities of the other one.

The PSF unit should include:

- PSF filters.
- The backwash system.
- The treated water pumping station.
- All the electromechanical and I&C equipment (pumps, piping, valves, instruments, etc.) necessary for proper operation.

4.9.1 Process Parameters

Table 4.14 summarizes the main process parameters for PSF filters.

Table 4.14 Main process parameters for PSF unit.

Parameter	Units	Value
Minimum filtering bed depth	m	1.2
Minimum height of cylindrical section	m	1.5
Maximum operative pressure drop	bar	0.8
Hydraulic load	m/h	9–12

4.9.2 Process Description

WW from the upstream treatment will be fed in FIC (centrifugal pump equipped with VFD) to the PSF filters. The PSF filter stage will allow further reduction of the residual TSS load present in the biological effluent.

Water to be treated will be applied to the top of the vertical bed cylindrical filter and withdrawn at the bottom. The sand bed will be held in place with an underlying drain system at the bottom of the column. The feed system should allow homogeneous feeding of the filtering bed through the sand media and water percolation through the bed should avoid the formation of preferred filtering paths and consequent "dead zones" within the filtration bed.

Filtered water from the PSF unit will then be stored in a dedicated tank and pumped by means of centrifugal pumps (one duty, one standby) to the following UF unit. PSF filtrate tank sizing should take into account the UF filtrate production pauses due to the backwashing and chemical cleaning cycles.

Decay of the filtration capacity of the sand bed due to its progressive saturation will be monitored by a differential pressure controller (if excess suspended matter is accumulated, the head loss through the filter increases until a maximum predetermined limit [1 bar] that will activate filter backwashing). However, automatic backwashing is normally activated once per day on a time-based logic (typically 15 minutes of backwashing).

As on option, a further control can be performed by an in-line TSS analyzer installed on the filtered water line. The sand bed will have to be backwashed when the TSS value and/or the pressure drop increases to the set high limit for the effluent quality.

Backwashing will allow the expansion of the sand layer so that the solids captured during normal operation can be released and separated from the sand. Backwashing water will be sent back to the equalization tank of the WWTP. The water used for the backwashing will be taken from the UF filtered water storage tank.

The sand filters configuration will be constituted by a minimum of two filters in parallel which will be set up and provided with automatic valves in order to allow automatic switch of the operating units order within the filtration cycle: in case one PSF filter is out of operation or in backwashing, a valve will divert the total flow to the other one to ensure continuous water filtration. This is the optional case. Normally one PSF will be in operation. The unit will have integral instrumentation for the proper automation through a dedicated PLC system.

4.10 Activated Carbon Filter

Activated carbon filter (ACF) polishing treatment is an optional tertiary treatment in WW treatment facilities. The ACF unit performs WW polishing through adsorption of soluble residual organic BOD/COD on ACF media. The ACF adsorption process should be performed as optional treatment upstream RO for water reuse (to avoid RO membrane fouling) only if BOD concentration downstream of the biological treatment exceeds 30 ppm.

To avoid clogging of the ACF due to TSS/turbidity, the ACF needs to be installed on the downstream side:

- If the chosen biological option is an MBR system, the ACF unit is installed directly downstream of the MBR, polishing the MBR permeate.
- If the chosen biological option is an MBBR or a CAS system, the ACF unit will be installed downstream of the PSF and UF units.

The ACF unit will be provided with a sparing philosophy considered to have, for each of the most critical rotating equipment, a 1 + 1 philosophy, with an installed spare that can be automatically switched into operation in case of failure of the main one, in order to ensure the continuity of the operation of the WWTP. Sparing one will be optional. Normally one ACF filter will be in operation.

Adsorption on activated carbon media is the heart of the process; contacting between WW and the ACF will be performed in a number of fixed-bed columns, organized to achieve an in-series contact process; the number of contact columns will depend on the plant size and should in any case include two carbon filters (as minimum), with the possibility to have each of them interchangeable as a first/second one in order to:

- Allow the complete saturation of the first contact column (and substitution of not completely saturated ACF), having the second one in series as a guard.
- Guarantee at least one operating filter during maintenance/carbon replacement activities of the other one.

The ACF unit should include:

- The ACF contact columns.
- The treated water pumping station.
- All the electromechanical and I&C equipment (pumps, piping, valves, instruments, etc.) necessary for proper operation.

4.10.1 Process Parameters

Table 4.15 summarizes the main process parameters for the ACF.

Table 4.15 Main process parameters for GAC filters.

Parameter	Units	Value
Carbon adsorption rate	kg COD/kg GAC	0.25
HRT	min	15
Maximum operative pressure drop	bar	0.8
Hydraulic load	m/h	9–10

4.10.2 Process Description

The WW inlet will be fed directly under pressure from the upstream treatment unit to the ACF filters. The ACF filter stage will allow the operator to further reduce the residual part of the organic matter present in the upstream treatment effluent. Water to be treated will be applied to the top of the column and withdrawn at the bottom. The carbon will be held in place with an underlying drain system at the bottom of the column.

The decay of the adsorption capacity of the carbon due to its progressive saturation will be monitored by an in-line TOC analyzer installed on the filtered water line. The carbon bed will have to be replaced when the TOC value increases to the set high limit for the effluent quality.

The ACF filter train will be set up and provided with automatic valves in order to allow automatic variation of the operating units order within the filtration cycle; this will allow the operator to organize as terminal element the filter that has been subjected to the most recent carbon charge renew. In case one ACF filter is out of operation, a valve will divert the total flow to the other one. ACF filtered water will be stored in a dedicated tank and sent to further downstream units.

Backwashing operations should be performed in order to limit the head loss buildup due to the removal of particulate suspended solids within the carbon column. The backwash phase will be manually initiated through head loss measurement. Filter backwashing may also be performed on a fixed time basis (typically once per day for 10 minutes). Backwash can be performed with either:

- filtrate water from a UF filtrate storage tank, if a UF unit is installed upstream; or
- net water, if no UF unit is installed upstream.

The organic matter content and the temperature of the influent will determine the carbon saturation limit and consequently the actual carbon consumption. The carbon will be periodically regenerated and replaced to ensure effective BOD removal. The unit will have integral instrumentation for proper automation through a dedicated PLC system.

4.11 Ultrafiltration

A UF unit is one of the tertiary treatments selected in WWTP. UF treatment has to be installed downstream of the PSF when upstream biological treatment is performed with CAS or MBBR. No UF treatment is needed when MBR is adopted as biological treatment.

The UF unit performs pressure-driven separation of particulate and microbial contaminants from the pretreated WW stream. The unit will be provided in one single line but with a sparing philosophy considered to have, for each of the most critical rotating equipment, a 1 + 1 philosophy, with an installed spare that can be automatically switched into operation in case of failure of the main one, in order to ensure the continuity of the operation of the WWTP. The UF unit should include:

- The UF membrane separation system.
- UF permeate water storage tank.
- Backwash pumping unit.
- Cleaning in place (CIP) unit.
- Blower unit for backwashing.
- Chemical storage and dosing systems for membranes cleaning (chemically enhanced backwash [CEB] or CIP).
- All the electromechanical and I&C equipment (pumps, compressor, mixers, instruments, etc.) necessary for proper operation.

4.11.1 Process Parameters

Table 4.16 summarizes the main process parameters for the UF unit.

Table 4.16 Main process parameters for UF unit.

Parameter	Units	Value
Molecular weight cut-off	μm	0.03
Membrane design-specific flux	$l/m^2/h$	≥ 60
Module diameter	m	0.225
Module length	m	1.5
Module volume	liters	35
UF membrane area	m^2	33

4.11.2 Process Description

Process WW from PSF is passed through UF.

- A suitable number of micro-strainers, 100 μm filtration grade, remove coarser materials in order to protect the UF membranes installed downstream.
- UF membranes installed in one train will remove particles greater than 0.03 μm, which should be sufficient to remove grazing microorganisms and potential biological contamination, if needed.

The UF system requires backwash, cleaning, and flushing cycles automatically managed by a dedicated control panel. During these cycles, WW will be produced and it will be sent back to the equalization tank. Since PSF filtered water fills in continuous mode the storage tank, it should be sized to account for the UF train service interruptions due to backwashing and CEB and CIP cycles (see sections below).

UF permeate is sent to a storage tank in which three pump sets are installed: one delivers the stream to the downstream treatment, the second one is dedicated to the backwash of the UF membranes, and the third one pumps the permeate water to the PSF unit for filter washing operations and to GAC filters (if installed downstream) for washing operations. Since the GAC and PSF units require different water flow rates depending on GAC and PSF filter volume, the latter pump set should be equipped with VFD controlled by an FIC or, alternatively, a flow control valve can be installed for flow regulation. Moreover, the UF permeate storage tank should be sized to account for the above-cited permeate uses and for permeate production interruptions due to backwashing, CEB, and CIP cycles.

4.11.2.1 Configuration and Operation

The following sections describe a typical UF configuration and operation mode, and include an overview of backwashing and cleaning cycles.

4.11.2.1.1 Membrane Configuration

The modules consist of hollow fiber, polyvinylidene fluoride (PVDF) polymer membranes which are installed in a pressurized vertical "shell-and-tube" design that eliminates the need for separate pressure vessels. The membrane operative configuration is outside-in, dead-end flow. Outside-in flow configuration is tolerant of wide ranging feed water qualities and allows air scour cleaning. Dead-end flow offers higher recovery and energy savings.

There are four connections on each UF module. The flow enters the module through the side port located on the bottom end cap. The air feed is located on the bottom of the end cap and is used for air scouring on the outside of the fiber during cleaning. The concentrate (discharges flow from outside of fiber) and permeate ports (inside of fiber) are located on the top cap.

4.11.2.1.2 Membrane Normal Operation

Normal operation of the UF system refers to the routine operating sequence of a system and includes the operating and backwash steps. At initial start-up, the modules are flushed using "forward flush" to remove any residual chemicals or trapped air from the module. The flush occurs on the outside of the fibers and does not filter the feed water. After the

Table 4.17 UF membranes typical operating conditions.

Parameter	Units	Value
Maximum operating feed pressure	bar	2.5
Maximum operating TMP	bar	1.5
Backwash flux	$l/m^2/h$	100–300
Permissible temperature range	°C	1–40

"forward flush," the modules can be placed in the operating mode. An operating cycle ranges from 20 to 60 minutes. While operating, 100% of the feed water is converted to permeate. As contaminants are removed during the operating phase, the transmembrane pressure (TMP) will rise. At the end of the preset operating cycle time, a backwash sequence is triggered in order to remove particulate fouling from the membranes. Moreover, a control system on TMP set at a specific set-point value activates the backwash procedure if the TMP exceeds the fixed limit value that can reduce system efficiency before the preset cycle time. UF membranes typical operating conditions are shown in Table 4.17.

4.11.2.1.3 Backwash

During backwash mode, a small quantity of the produced permeate is pumped back through the membranes with a backwash pump (one duty, one standby). In this process the fouling layer at the surface of the membrane caused by particulate matter is disrupted and the pores are cleaned. The frequency and the duration of the backwash are adjustable for the system.

The backwash mode may include an air scour, and always includes draining, backwash through the top drain, backwash through the bottom drain, and a forward flush.

Water backwash is operated in 24 cycles/day, 1 minute/cycle. The main steps of backwashing are as follows:

- The air scour step, when included, is used to loosen particulates deposited on the outside of the membrane surface. Air is introduced on the outside of the fibers and displaced feed flow/concentrate is allowed to discharge through the top of the module for disposal. After 20–30 seconds of air scour the module is drained by gravity to remove dislodged particulates.
- If the air flush is not included, the backwash sequence is started by simply draining the module by gravity to remove the concentrated feed water before starting any backwashing.
- After draining, the first backwash step is performed. Permeate flow is reversed from the inside of the fiber to the outside and backwash flow is removed from the module housing through the top drain on the module. A top draining backwash is performed first to remove contaminants in the area of greatest concentration.
- The second backwash step is performed to remove contaminants through the bottom of the module housing. Permeate flow is reversed from the inside of the fiber to the outside and backwash flow is removed from the module housing through the bottom drain on the module for efficient removal of heavier materials. The two steps of backwash can be repeated numerous times depending on the degree of fouling.

- After backwash is complete, a forward flush is performed to remove any remaining contaminants and remove any air trapped on the outside of the fibers. After backwash, the modules are returned to the operating mode.

More detailed procedures for operating conditions of the membranes will be specified by the selected vendor.

4.11.2.1.4 Membrane Shutdown

The backwash mode of the membranes should be operated also each time one module is shut down, in order to prevent the membrane irreversibly losing flux because of dry-out. Indeed, UF systems are designed to run continuously and membrane systems perform better when operated continuously. When the UF train is shut down, the system must be cleaned using air-scour and backwashed with permeate water to prevent bacterial growth in the UF system.

The water used for backwash before shutdown should not contain chemicals. Any feed water and backwash chemical dosing used should be stopped before the last cleaning and shutdown. After cleaning, all valves on the UF system should be closed to seal the system.

In general, if the modules are shut down for more than 48 hours, preservatives like sodium bisulfite should be added to the backwash flow. Table 4.18 summarizes the recommended steps for membrane maintenance for different shutdown durations. Detailed procedures for shutdown of the membranes will be specified by the selected vendor.

Table 4.18 UF skids shutdown summary.

Storage duration	Recommended steps
0–48 hours	Air scour and backwash, close all valves
>2 days to 7 days	Air scour and backwash, close all valves, perform 30–60 minutes of operations daily or air scour, backwash, add preservative, then close all valves
>7 days to 90 days	Add preservative and renew preservative every three months
>90 days	Consult manufacturer

4.11.2.1.5 Chemically Enhanced Backwash

CEB operations are quick-maintenance cleaning procedures used to remove biological, organic, and inorganic fouling from the membranes. CEB has the aim to extend the operating times between major cleanings (CIP) and is performed with a frequency dependent on the feed water quality. On high-quality feed waters a CEB may not be required since simple backwash is enough to clean the membranes. The CEB process occurs automatically, but the frequency can be adjusted after gaining operating experience. CEB is performed using UF permeate, and an acid or base combined with an oxidant is added to more effectively clean contaminants from the membrane surface.

The following are common CEB chemicals (to be confirmed by the vendor):

- An oxidant – sodium hypochlorite (NaOCl) for removal of biological fouling.
- An oxidant and base – sodium hypochlorite (NaOCl) and sodium hydroxide (NaOH) for removal of organic fouling.
- An acid – hydrochloric acid (HCl) or citric acid for removal of inorganic fouling.

CEB follows the steps of a normal backwash with the following differences:

- A chemical is dosed into the backwash water.
- A soak step is added after the second backwash step. The soak is performed for 5–20 minutes in order to allow time for the chemical to react with contaminants that have attached to the membrane surface or penetrated the fiber wall. Intermittent air scour can be applied during the soak step.

After the soak a routine backwash is performed to remove any remaining particulates and residual chemicals. Detailed procedures for CEB cleaning of the membranes, chemicals, concentrations, and dosages will be specified by the selected membrane vendor.

4.11.2.1.6 Cleaning in Place

A CIP operation is a complete cleaning process which includes backwash and chemical recycling to clean the fibers. CIP is an "on demand" operation and is performed manually by the operator. CIP consists of an extensive chemical soak and is recommended when the TMP cannot be lowered by either simple backwash or CEB operations. The frequency of CIP is dependent on the feed water quality but can range from one to three months.

The chemicals used for CIP are usually the same used for CEB (to be confirmed by the vendor):

- An oxidant – sodium hypochlorite (NaOCl) for removal of biological fouling.
- An oxidant and base – sodium hypochlorite (NaOCl) and sodium hydroxide (NaOH) for removal of organic fouling.
- An acid – hydrochloric acid (HCl) or citric acid for removal of inorganic fouling.

Compared to CEB, for CIP operations HCl and NaOH are dosed at higher concentrations.

Detailed procedures for recovery cleaning of the membranes, chemicals, concentrations, and dosages will be specified by the selected membrane vendor.

The typical CIP cycle is as follows:

- Prior to CIP, the routine backwash steps are performed. The backwash steps are repeated three to eight times to remove contaminants or foulants not requiring chemical removal.
- After completing the backwash steps, the module is drained by gravity to remove excess water and prevent dilution of the CIP chemicals.
- CIP chemicals are recycled on the outside of the module for 30 minutes through the CIP tank. A small chemical permeate stream will also be collected and recycled to the CIP tank.
- A soak follows the initial recycle step for 60 minutes or longer depending on the degree of fouling that has occurred.
- After the soak step, CIP chemicals are again recycled on the outside of the module for 30 minutes.
- When the recycle is completed an air scour is performed and then the module is drained to remove concentrated chemicals.
- The two steps of backwash and a forward flush are performed to remove any remaining contaminants on the outside of the fibers.

After CIP and at the start of the operating step, permeate may be used to remove residual chemicals held in the fiber or module.

The CIP steps described above are for a single alkali or acid chemical solution. If acid and alkali cleaning are required, the CIP steps would be repeated for each chemical solution. Detailed procedures for recovery cleaning of the membranes, chemicals, concentrations, and dosages will be specified by the selected membrane vendor.

4.12   Reverse Osmosis

An RO treatment unit is part of the tertiary treatments of WWTPs used to reduce dissolved salts from treated WW. The RO unit may include:

- a single-stage RO system; or
- a double-stage system, consisting of two RO subunits that perform the following operations in series:
  - a first RO stage for the treatment of the WW coming from the upstream treatments;
  - a second RO stage for the treatment of the brine produced in the first RO stage.

The latter option should be considered if there is the need to maximize water reuse, and a further concentration of the brine generated by the first RO stage is required in order to minimize the brine stream to be disposed of.

The RO unit will be provided in one single line but with a sparing philosophy considered to have, for each of the most critical rotating equipment, a $1+1$ philosophy, with an installed spare that can be automatically switched into operation in case of failure of the main one, in order to ensure the continuity of the operation of the WWTP. The RO unit should include:

- One cartridge filter of 5 μm.
- High pressure pumping unit.
- First-stage RO treatment system.
- Second-stage RO treatment system (optional).
- Antiscalant dosage system.
- Sodium metabisulfite dosage system.
- Acid dosing system (hydrochloric/sulfuric acid).
- Permeate/CIP buffer tank.
- Permeate/CIP pumping system.
- All the electromechanical and I&C equipment (pumps, instruments, etc.) necessary for proper operation.

4.12.1   Process Parameters

Table 4.19 summarizes the main process parameters for the RO unit. The dimensions of each RO membrane would be 8 x 40 inches, and RO vessels will contain six RO membranes.

Table 4.19   Main process parameters for RO unit.

Parameter	Units	Value
Module diameter	m	0.2
Module length	m	1.0

4.12.2 Process Description

The unit consists of a single-stage RO system or of a double-stage system depending on the chosen option for the treatment of the first RO brine stream. The second-stage RO treatment option should be considered if there is the need to maximize water reuse and further concentration of the brine generated by the first RO stage in order to minimize the brine stream to be disposed. Upstream of the first-stage RO membranes the following products will be used:

- Antiscalant as scale inhibitor.
- Sodium metabisulfite to reduce oxidizing agents present in WW (since a chlorination basin is installed upstream of the RO treatment). An oxidation reduction potential (ORP) control should be installed in-line to control the sodium metabisulfite dosage.
- Acid dosing for reduction of biofouling.

A single stage of filtration cartridge acts as a safety barrier to protect the membranes from TSS fouling and a 5-µm cartridge filter is installed for this purpose. The system should be provided with an automatic flushing and manual CIP unit.

The RO unit will therefore include a permeate buffer tank to store permeate for cleaning purposes (flushing and CIP). Every time the plant stops, a flushing will be done, displacing the brine from the membranes with permeate and avoiding potential scaling.

When necessary, CIP provides in-line cleaning to the RO vessels to restore their operating conditions. CIP chemicals are prepared in the same permeate buffer tank and dosed upstream of the first-stage RO vessels. Generally, CIP is done not more than four times a year (once every three months).

4.12.2.1 First-Stage RO Treatment

The inlet WW coming from previous sections is fed by means of high-pressure centrifugal pump (one duty, one spare) to the first-stage RO unit membranes. The permeate from the first-stage RO unit is sent to the treated water storage tank. The brine generated from the first-stage RO can be:

- Directly discharged to sea, brackish water, or river (if permitted by local regulations).
- Blended with fresh/treated water and reused for gardening or toilet flushing.
- Sent to an evaporation pond (for warm climates).
- Treated with a second RO stage and subsequently treated with an evaporator followed by a crystallizer.

The first-stage RO membrane will be 8 inches in diameter by 40 inches in length. Table 4.20 shows the characteristics of first-stage RO membranes. Therefore the calculation should be performed as follows:

- RO permeate flow (m^3/h) = Inlet flow rate (m^3/h) x RO recovery (70%);
- Number of membranes (n) required = RO permeate flow (m^3/h)/Design-specific flux $(15 l/m^2/h)$/Membrane surface area $(37 m^2/membrane)$ x 1000 (in normal operation);
- Number of vessels required (n vessels) = Number of elements required (n) / Number of elements per each vessel (6 membranes/skid).

Table 4.20 Characteristics of First Stage RO Membranes.

Parameter	Units	Value
Design-specific flow for RO (with all skids in operation)	$l/m^2/h$	17–20
RO membrane surface	m^2	37
Membranes contained in one RO vessel	n	6

4.12.2.2 Second-Stage RO Treatment

The second-stage RO unit is fed with the brine produced from the first-stage RO treatment, in order to obtain concentrated RO brine, enhancing the overall desalination treatment process. The permeate streams from both RO stages go into the common permeate storage tank which receives also the overhead from the evaporator unit. The brine produced from the second-stage unit is sent to the evaporator with crystallizer unit where a further concentration of the brine will be performed. The second-stage RO membrane will be 8 inches in diameter by 40 inches in length. Table 4.21 shows the characteristics of second-stage RO membranes. Therefore the calculation should be performed as follows:

- Second-stage RO permeate flow (m^3/h) = first-stage RO brine flow rate (m^3/h) × second-stage RO recovery (55%).
- Number of membranes (n) required = Permeate flow (m^3/h)/Design-specific flux $(15 l/m^2/h)$/Membrane surface area $(37 \ m^2$/membrane) × 1000 (in normal operation).
- Number of vessels required (n vessels) = Number of elements required (n)/Number of elements per each vessel (6 membranes/skid).

The total volume of RO vessels (first stage + second stage), useful to determine the flushing and chemical cleaning volume and therefore the permeate/CIP tank volume, is calculated as:

$$\text{Total RO modules volume} \left(m^3\right) = \text{Membrane volume} \left(0.03 \, m^3\right) \times$$
$$\left(\text{number of first} - \text{stage RO membranes}\left[n\right] + \text{number of second} - \text{stage RO membranes}\left[n\right]\right)$$

Table 4.21 Characteristics of Second Stage RO Membranes.

Parameter	Units	Value
Design-specific flow for RO (with all skids in operation)	$l/m^2/h$	15
RO membrane surface	m^2	37
Membranes contained in one RO vessel	n	6

4.12.3 Membrane Cleaning

RO membranes are widely used for salty water treatment. The use of RO membranes in water desalination and WW reclamation, reuse, and recycling has increased over the past few years. However, a major impediment is membrane fouling. During normal operation over a period of time, RO membrane elements are subject to fouling by suspended, organic, or inorganic materials that may be present in the feed water. Common examples of foulants are:

- Calcium carbonate scale.
- Sulfate scale of calcium, barium, or strontium.
- Metal oxides (iron, manganese, copper, nickel, aluminum, etc.).
- Polymerized silica scale.
- Inorganic colloidal deposits.
- Mixed inorganic/organic colloidal deposits.
- Natural organic matter (NOM).
- Man-made organic material (e.g. antiscalant/dispersants, cationic polyelectrolytes).
- Biological matter (bacterial bioslime, algae, mold, or fungi).

The entry of these foulants in RO systems causes fouling of membranes. The fouling leads to increase in the differential pressure from feed to concentrate and finally leads to membrane flux declination and mechanical damage of the membrane. Foulants removal through chemical cleaning is therefore a major objective of the membrane chemical cleaning process.

4.12.3.1 Chemical Cleaning Requirement

Membranes of RO systems must be cleaned with suitable chemicals in any of the following circumstances:

- When the normalized permeate flow drops by 10%.
- When the normalized salt passage increases by 5%.
- When the normalized differential pressure increases by 15%.

RO chemical cleaning frequency will vary by site. A rough rule of thumb as to an acceptable cleaning frequency is once every three to five months. It is important to clean the membranes when they are only lightly fouled, not heavily fouled.

4.12.3.2 Chemical Cleaning Sequence

Normally three kinds of RO chemical cleaning are recommended: acid cleaning, alkaline cleaning, and sanitizing. The recommended sequence is given below:

- Acid cleaning and flushing.
- Alkaline cleaning and flushing.
- Sanitizing and flushing.

4.12.3.3 Cleaning Chemicals

Acid cleaners and alkaline cleaners are the two standard cleaning chemicals. Acid cleaners are used to remove inorganic precipitates including iron, while alkaline cleaners are used to remove organic fouling including biological matter. It is recommended that RO permeate should be used for the cleaning solutions. Table 4.22 lists suitable RO cleaning chemicals along with recommended concentrations.

Table 4.22 Recommended RO cleaning chemicals.

Chemical	Concentration by weight (%)	Foulant
Sodium hydroxide (NaOH)	0.1	Biological and organic
Tetrasodium salt of ethylene Diamine tetra-acetic acid (Na_4EDTA)	1.0	Biological and organic
Sodium salt of dodecylsulfate (Na-DDS)	0.025	Biological, organic, and silica
Hydrochloric acid (HCl)	0.2	Inorganic
Sodium hydrosulfite ($Na_2S_2O_4$)	1.0	Inorganic and iron
Phosphoric acid (H_3PO_4)	0.5	Inorganic and iron
Sulfamic acid (NH_2SO_3H)	1.0	Metal oxides
Hydrogen peroxide	0.25	Bacterial (*E. coli*)

4.12.3.4 Flow and Pressure of Chemical Solutions

Flow rate and pressure of circulated chemical solutions during chemical cleaning of RO membranes play a very important role. Table 4.23 lists required flow and pressure during chemical cleaning of RO membranes.

Table 4.23 Recommended flow and pressure of RO chemical cleaning solution.

Membrane diameter (inches)	Flow rate per pressure vessel (m^3/h)	Pressure (bar)
2.5	0.7–1.2	1.5–3.5
4	1.8–2.3	1.5–3.5
8	6.0–9.0	1.5–3.5

4.12.3.5 Temperature and pH Range of Chemical Solutions

Temperature and pH of circulated chemical solutions during chemical cleaning of RO membranes also play a very important role. Table 4.24 lists required temperature and pH range during chemical cleaning of RO membranes.

Table 4.24 Recommended temperature and pH range of RO chemical cleaning solution.

Temperature (°F)	Operating pH range	Cleaning pH range
20–120	3–11	2–12

4.12.3.6 Chemical Cleaning Equipment

RO chemical cleaning systems consist of a mixing tank, cleaning pump, and micron cartridges. The successful cleaning of an RO on-site requires well-designed RO cleaning equipment. It is recommended to clean a multistage RO one stage at a time to optimize

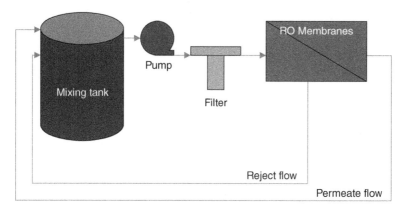

Figure 4.6 General arrangement of RO chemical cleaning equipment.

Table 4.25 Size and material of constructions of RO chemical cleaning equipment.

Equipment	Size	Material of construction
Mixing tank	20% more to the volume of pressure vessel	HDPE/PP/FRP
Cleaning pump	Flow	SS-316/PP
Cartridges filters	10 µm	

HDPE, High-density polyethylene; FRP, fiber reinforced polymer; SS, stainless steel; PP, polypropylene.

cross-flow cleaning velocity. The source water for chemical solution make-up and rinsing should be clean RO permeate or deionized (DI) water and be free of hardness, transition metals (e.g. iron), and chlorine. Components must be corrosion proof. The general arrangement of RO chemical cleaning equipment is shown in Figure 4.6. Table 4.25 lists size and material of constructions for RO chemical cleaning equipment.

4.12.3.7 Chemical Cleaning and Flushing Procedures
In the RO plant the RO membrane can be cleaned in place in the pressure tubes by recirculating the cleaning solution across the high-pressure side of the membrane at low pressure and relatively high flow.

- Flush the pressure tubes at low pressure (3.5 bar or 50 psi) by pumping clean water (RO permeate or DI quality and free of hardness, transition metals, and chlorine) from the cleaning tank.
- Mix a fresh batch of the selected cleaning solution in the cleaning tank. The dilution water should be clean water of RO permeate or DI quality and be free of hardness, transition metals, and chlorine. The temperature and pH should be adjusted to their target levels.
- Circulate the cleaning solution through the pressure tubes for approximately one hour or the desired period of time. At the start, send the displaced water to drain so you do not dilute the cleaning chemical, and then divert up to 20% of the most highly fouled cleaning solution to drain before returning the cleaning solution back to the RO cleaning tank. For the first five minutes, slowly throttle the flow rate to one-third of the maximum

design flow rate. This is to minimize the potential plugging of the feed path with a large amount of dislodged foulant. For the second five minutes, increase the flow rate to two-thirds of the maximum design flow rate, and then increase the flow rate to the maximum design flow rate. If required, readjust the pH back to the target when it changes more than 0.5 pH units.

- An optional soak and recirculation sequence can be used, if required. The soak time can be from one to eight hours depending on the manufacturer's recommendations. Caution should be used to maintain the proper temperature and pH. Also note that this does increase the chemical exposure time of the membrane.
- Upon completion of the chemical cleaning steps, a low-pressure cleaning rinse with clean water (RO permeate or DI quality and free of hardness, transition metals, and chlorine) is required to remove all traces of chemical from the cleaning skid and the RO skid. Drain and flush the cleaning tank; then completely refill the cleaning tank with clean water for the cleaning rinse. Rinse the pressure tubes by pumping all of the rinse water from the cleaning tank through the pressure tubes to drain. A second cleaning can be started at this point, if required.
- Once the RO system is fully rinsed of cleaning chemical with clean water from the cleaning tank, a final low-pressure clean-up flush can be performed using pretreated feed water. The permeate line should remain open to drain. Feed pressure should be less than 60 psi (4 bar). This final flush continues until the flush water flows clean and is free of any foam or residues of cleaning agents. This usually takes 15–60 minutes. The operator can sample the flush water going to the drain for detergent removal and lack of foaming by using a clear flask and shaking it. A conductivity meter can be used to test for removal of cleaning chemicals, such that the flush water to drain is within 10–20% of the feed water conductivity. A pH meter can also be used to compare the flush water to drain to the feed pH.

1) Once all the stages of a train are cleaned, and the chemicals flushed out, the RO can be restarted and placed into a service rinse. The RO permeate should be diverted to drain until it meets the quality requirements of the process (e.g. conductivity, pH).

4.13 Evaporator with Crystallizer

An evaporator with crystallizer unit is one alternative treatment for brine management within the WW treatment facilities. Such a unit treats the brine produced in the second-stage RO treatment and other high TDS WW not able to be treated with biological processes.

The unit will be provided with one treatment line unless specific plant size requires more than one line to be installed. A sparing philosophy will be adopted considered as having, for each of the most critical rotating equipment, a 1 + 1 philosophy, with an installed spare that can be automatically switched into operation in case of failure of the main one, in order to ensure the continuity of the operation of the WWTP. The unit should include:

- Atmospheric thin film evaporator (or tubular falling film-type evaporator).
- Vapor compression crystallizer (optional).
- Preheating heat exchangers for both units.
- Distillate storage tanks.

- Distillate pumps.
- Concentrate storage tanks.
- Concentrate pumps.
- Chemicals storage and dosage for chemical cleaning.
- All the electromechanical and I&C equipment (pumps, piping, valves, compressor, mixers, instruments, etc.) necessary for proper operation.

4.13.1 Process Parameters

Table 4.26 summarizes the main process parameters for the evaporator with crystallizer.

Table 4.26 Main process parameters for evaporator with crystallizer.

Parameter	Units	Value	Notes
TDS concentration in overhead	mg/l	100	Overhead stream has to be blended with RO permeate and should be suitable for reuse within the facility
Specific heat requirement (assuming feed at 25 °C)	kW/m^3 feed	90	

4.13.2 Process Description

4.13.2.1 Process Waste Water Preheating

The brine coming from the second-stage RO unit will be preheated in a battery of heat exchangers in which the evaporation overhead will be fed in countercurrent. During this phase the evaporation overhead will release part of its thermal energy to the incoming brine, allowing an energy saving in the overall process. The operation temperature maintained inside the evaporator is around 90 °C, and so – considering some thermal losses – the overhead will leave the evaporators at a temperature of around 80 °C. The energy recovery will depend on the heat exchanger sizing. Depending on the plant size a further energy saving can be achieved by including a secondary heat exchange battery in order to recover the thermal energy of the concentrated brine; this flow will be lower in terms of flow rate compared with the overhead but also its temperature will be around 80 °C at the outlet from the evaporator.

4.13.2.2 Evaporator

The second-stage RO brine, after preheating, will be fed to a mechanical falling film evaporator (or a thin film type), provided with a mechanical vapor recompression system to recover latent heat of evaporation. The following description applies for a mechanical falling film evaporator with mechanical vapor recompression.

The operation of the evaporation unit is based on the mechanical compression evaporation process, which allows aqueous components to be evaporated, while non-volatile components are retained in the concentrate. The produced overhead is virtually free from TDS (<100 mg/l) and can be sent to the treated water storage together with the RO permeate. Filling of the unit will be driven by the vacuum created by the compressor.

The main steps of the evaporation are as follows:

- The preheated brine flow fed to the evaporator sump will be circulated through heat transfer tubes internal to the machine where the outcoming overhead flow releases heat to the brine.
- Part of the brine evaporates, while the remaining brine flows in a falling film into the brine sump and is then recirculated by a centrifugal pump inside the tubes; part of the brine is blown down.
- The mechanical compressor will generate a negative pressure on the suction side where the vapor will be withdrawn and will increase the vapor pressure and temperature on the delivery side.
- The high-temperature vapor will be delivered by the compressor to the main vessel heat exchanger, where it will condensate, releasing thermal energy to the evaporating brine falling inside the tubes, and then it will leave the evaporator as distillate.
- The distillate is pumped back to the preheater where it releases part of the heat to the incoming brine flow.

The condensation of the overhead will allow the establishment of a natural circulation of the evaporating fluid between the suction side and the delivery side of the compressor due to difference in density. Before the compression stage the vapor will be purified by means of mist eliminators.

During the evaporation phase the distillate will leave the machine continuously; when the setup conditions are attained, the evaporation phase will be over, and the concentrated brine will be delivered outside the machine. At that point a new heating phase will take place and the process can restart. For plants having a size like the ones of Unilever, evaporators are not usually operated continuously 24 h/day but are operated in batch mode. The contractor should in any case verify the best solution according to the specific case.

The heating phase duration is around 40 minutes, while the evaporation phase duration can be largely variable: the more the influent WW will be concentrated, the less the evaporation phase will last.

The distillate is collected in a pumping pit and sent to the treated water storage tank, while the brine is collected in a buffer tank provided with mixer and can then be sent:

- To crystallizer, if further brine concentration is required; or
- To evaporation ponds, if available; and
- To offsite disposal.

During evaporator operation, vents will be withdrawn from the machine and sent to atmosphere; depending on the plant size, a further heat exchanger battery can be considered in order to condensate part of the vents and recover further thermal energy from the process. The vent condensate will be mixed with the overhead. The unit will have integral level controls and instrumentation for the proper automation through a dedicated PLC system.

4.13.2.2.1 *Evaporator Cleaning*
The evaporator operation will require chemical washing cycles to remove scaling from heating surfaces. Generally, only an acid washing cycle is considered for plant performance maintenance; however, the vendor will specify the chemicals, frequencies, and dosages specific for the selected unit.

4.13.2.3 Crystallizer (Optional)

The concentrated brine from the evaporator may be sent to a vapor compression crystallizer in order to enhance brine concentration if there is the need to recover further distillate for reuse, and obtain a smaller flow of very concentrated brine to dispose offsite. The process of the vapor compression crystallizer is summarized as follows:

- The brine from the evaporator is fed to the crystallizer where it is preheated in a shell and tube heat exchanger where the vapor leaving the crystallizer releases heat to the influent brine stream; the brine is maintained under pressure, avoiding evaporation of the brine and preventing scaling formation on the tubes.
- The preheated brine enters the crystallizer with a specific angle in a way that it creates a vortex when entering the crystallizer; here, some of the brine evaporates, promoting the formation of crystals from the remaining brine.
- The brine is partially blown down and the remaining is recirculated to the preheater, while the overhead vapor is passed through a mist eliminator and then is compressed by a compressor.
- The compressed vapor exchanges heat with the influent brine stream in the shell and tube preheater and condenses in distillate.

The distillate is collected in a pumping pit and pumped to the treated water storage tank, while the brine is stored in a buffer tank provided with mixer and can then be sent to evaporation ponds, if available, and to offsite disposal.

4.14 Filter Press

A filter press (FP) is used for sludge dewatering. The other option considered for sludge dewatering is sludge drying beds, which may be considered only for warm and dry climates and if high surfaces are available.

The sludge and oil streams produced in the WWTP will be stored in the sludge/oil common storage tank; sludge thickening (may be considered as optional) will not be considered as a standard process given that both sludge and oil streams are collected in the same tank. The sludge is then pumped from the storage tank to the following dewatering unit, which includes a sludge dewatering filter press that will perform the solid/liquid separation; the dewatered slurry will be sent to optional further treatment (conditioning) and then to reuse or disposal; on the other hand, the separated supernatant will be recirculated to the equalization unit to be reprocessed in the WWTP.

The filter press unit will be provided in one single line but with a sparing philosophy considered to have, for each of the most critical rotating equipment, a $1 + 1$ philosophy, with an installed spare that can be automatically switched into operation in case of failure of the main one, in order to ensure continuity of the operation of the WWTP. The unit should include:

- Sludge/oil storage tank.
- Sludge/oil aeration system (piping and perforated pipes).
- Sludge/oil aeration blowers.
- Sludge/oil feed pumps.

- Plate and frame filter press.
- Filtrate storage tank.
- Filtrate pumps.
- Air scour blowers.
- Screw conveyors.
- All the electromechanical and I&C equipment (piping, valves, instruments, etc.) necessary for proper operation.

4.14.1 Process Parameters

Table 4.27 summarizes the main process parameters for the filter press dewatering unit.

Table 4.27 Main process parameters for filter press unit.

Parameter	Units	Value
TSS concentration in filtrate	mg TSS/l	100
TSS concentration in dewatered sludge	%	25

4.14.2 Process Description

4.14.2.1 Aerated Storage Tank

The sludge and oil streams incoming to the dewatering unit will be fed to a single aerated storage tank; this tank will allow a three- to four-day storage time in order to allow the sludge dewatering and conditioning process to be operated only 40 h/week (8 h/day, 5 days/week).

In order to avoid the formation of septic conditions within the tank and consequent odor problems, and to avoid solids settling, the storage tank will be equipped with an aeration system. The aeration will be provided with a dedicated lobe compressor and the relative piping. Inside the tank an appropriate number of perforated pipes (with 10-mm-diameter holes) will provide the oxygen transfer to the sludge.

Level instruments and alarms will be included for tank level control. The sludge will be withdrawn from the tank and fed to downstream treatments by screw pumps (one duty, one standby), also operated at variable speed by a VFD controlled by the filter press logics. The pump on duty will feed the filter press. This will allow the flow rate fed to the filter press to be changed following the operation rates of the machine.

4.14.2.2 Dewatering Unit – Plate and Frame Filter Press

The plate and frame filter press will receive the sludge and oil stream from the aerated storage tank through screw pumps. Before entering the machine the sludge will be eventually conditioned with an organic polymer that will enhance the filtration process; the dosage of this chemical will be performed with an in-line connection. The type of polymer and the operative dosage need to be adjusted with specific jar tests.

The filter press will be of fixed-volume, recessed-plate type; it consists of a series of rectangular plates, recessed on both sides, which are supported face to face in a vertical position on a frame with a fixed and movable head. A filter cloth is hung or fitted over each plate. The plates are held together with sufficient force to seal them to withstand the pressure applied during filtration. Hydraulic rams are provided to hold the plates together.

This filtration technology will allow treating effectively slurries containing oil independent of their settling propensity.

The filter press operation will be batched and divided into cycles, each of them including the following seven phases:

1) Filter closing.
2) Chambers filling.
3) Retention of the press under pressure causing slurry filtration.
4) Plate pack opening.
5) Plates shifting.
6) Cake detachment from the plates and consequent collection of the cake in the conveying hopper placed beneath the filter press.
7) Slurry feed channel air scour.

After those operations the filter press is ready for a new cycle. The indicative duration of the cycle is 30–50 minutes, depending on the characteristics of the sludge.

After filtration, dewatered slurry will be discharged from the panels onto a conveying screw hopper, and from there to an inclined screw conveyor. The screw will break the slurry cake and convey the material to the (eventual) downstream sludge conditioning unit. The capacity of the conveying hopper will be equivalent to the capacity of the filter press in one cycle. If sludge conditioning is provided, the hopper will have also a buffer function, storing the dewatered slurry for regular feed to the downstream units. The screw motors will be equipped with VFD for regular operation and soft starting.

For plants in which no sludge conditioning will be provided, the filter press can be installed on a raised structure above a dewatered sludge collection container that will be periodically emptied. The filtrate will be collected in a dedicated pumping pit and sent back to the equalization unit by means of centrifugal pumps (one duty, one spare). The filter press will be equipped with a scour blower (one duty, one spare); this equipment will be operated at the end of each cycle for the air scouring of the feed channel, to allow cake detachment. As an alternative, a service air net can be used for this purpose if it can guarantee the pressure requirement of the equipment.

All the operation phases will be fully automated and difference in the details will depend on the selected contractor. An automatic control should be installed to ensure that the filter press stops operations in case the slurry feed pump turns off. The unit will have integral level controls and instrumentation for proper automation through a dedicated PLC system included in the unit.

4.14.2.2.1 Periodical Washing

A periodical automatic high-pressure washing cycle with clean water should be performed indicatively every three to ten working days, this also depending on the filterability characteristics of the sludge. To avoid clean water usage, treated water from the WWTP will be used. The wash water will be collected in the filtrate storage tank and sent back to the equalization unit.

4.15 Belt Press

A belt press (BP) (or belt filter press) performs sludge dewatering by squeezing the sludge between two filter belts which go through a system of rolls, introducing shear and compression forces in order to separate the water from the sludge, thereby reducing sludge moisture content. This filtration technology allows effectively treating sludges independently from their settling capability.

The dewatered sludge can be sent for further treatment for reusing or simply being disposed of off-site; the separated liquid will be instead discharged to the WW sewer net and returned to the equalization tank.

The belt press unit will be provided in one single line but with a sparing philosophy considered to have, for each of the most critical rotating equipment, a 1 + 1 philosophy with an installed spare that can be automatically switched into operation in case of failure of the main one, in order to ensure the continuity of the operation of the WWTP. The belt press unit should include:

- Sludge storage tank (like a gravity thickener).
- Sludge aeration system (piping and perforated pipes).
- Sludge feed pumps (progressive cavity pumps – one duty, one standby).
- Belt press.
- Screw conveyors for dewatered sludge.
- Dewatered sludge storage roll-off bin.
- Polymer solution preparation tank.
- Polymer solution dosing pumps.
- All the electromechanical and I&C equipment (piping, valves, instruments, etc.) necessary for proper operation.

4.15.1 Process Parameters

Table 4.28 summarizes the main process parameters for the belt press dewatering unit.

Table 4.28 Main process parameters for belt press unit.

Parameter	Units	Value
TSS concentration in filtrate	mg TSS/l	100–300
Dryness of dewatered sludge	%	
Primary sludge		25–30
Biological sludge		12–20
Mixed primary/biological sludge		20–28
Belt speed	m/h	50–150
Loading per meter	kg TSS/m/h	
Primary sludge		360–550
Biological sludge		45–180
Mixed primary/biological sludge		180–320

4.15.2 Process Description

The sludge streams from upstream units will be stocked in a dedicated storage tank (potentially a gravity thickener) or fed directly from the producing units to the belt press.

The sludge will be fed to the belt press by means of screw pumps (one duty, one standby), equipped with a VFD controlled by the belt press logics.

Before entering the machine the sludge will be conditioned with an organic polymer that will enhance the dewatering process. The type of polymer and the operative dosage need to be adjusted with specific jar tests. The dewatering belt press is intended to produce dewatered cake with a high solids content. Conditioned sludge will first enter a gravity drainage stage to remove free-draining water.

Following the gravity section, the sludge will enter a low-pressure section, with the top belt being solid and the bottom belt being a sieve where further water removal occurs. As a last step, the sludge will go through a high-pressure zone of belts with a serpentine or sinusoidal configuration to increase shears and pressures. Basically, belts are fed through the rollers and water is squeezed out of the sludge. When the belts pass through the final pair of rollers in the process, the filter cloths are separated and the filter cake is scraped off on an inclined screw conveyor and dropped into a roll-off bin for further treatment or disposal. The filtered liquid will be discharged into a centrate chute that is piped to the WW sewer net.

A pneumatically adjustable belt tensioning system will allow adjustment while the press is in operation. The unit will have integral level controls and instrumentation for proper automation through a dedicated PLC system included in the unit.

A periodical automatic high-pressure total washing cycle with clean water should be performed indicatively every three to ten working days, this also depending on the characteristics of the sludge. Water for cleaning sprays will be filtered plant effluent or utility water with a TSS level of 10–20 mg/l. The water pressure supplied to the belt press should be a minimum of 8 bar (120 psi). The wash water will be discharged to the WW sewer net.

4.16 Centrifuge

A centrifuge unit is an equipment for continuous sludge dewatering. The centrifuge is fed with sludge stored in the sludge storage tank or in the sludge thickener (in some applications it could be fed directly from the biological unit). The sludge enters to the rotating bowl of the centrifuges, where, by centrifugal force, it is separated into a concentrated cake and a dilute stream called "centrate." The cake is then discharged by a screw feeder into a screw conveyor and then is sent to reuse or final disposal. The centrate is usually recirculated in the equalization tank or into the biological system.

The centrifuge unit will be provided in one single line but with a sparing philosophy considered to have a bypass system that can be automatically switched into operation in case of failure of the centrifuge, in order to ensure the continuity of the operation of the WWTP. The unit should include:

- Sludge storage tank.
- Sludge feed pumps.

- Sludge aeration system (piping and perforated pipes).
- Sludge aeration blowers.
- Sludge centrifuge decanter.
- Screw conveyors.
- Dewatered sludge storage tank.
- Polymer solution preparation tank.
- Polymer solution dosing pumps.
- All the electromechanical and I&C equipment (piping, valves, instruments, etc.) necessary for proper operation.

4.16.1 Process Parameters

Table 4.29 summarizes the main process parameters for the centrifuge dewatering unit.

Table 4.29 Main process parameters for centrifuge unit.

Parameter	Units	Value
TSS concentration in centrate	mg TSS/l	100
Dryness of dewatered sludge	%	
Primary sludge		25–35
Biological sludge		14–18
Mixed primary/biological sludge		15–25

4.16.2 Process Description

The centrifuge will receive the sludge and oil stream from the sludge storage tank (or gravity thickener) through progressive cavity feed pumps. Before entering the machine the sludge will be conditioned with an organic polymer that will enhance the dewatering process; the dosage of this chemical will be performed with an in-line connection. The type of polymer and the operative dosage need to be adjusted with specific jar tests.

The centrifuge should be continuously fed with sludge conditioned with polymer by individual sludge feed pumps. The sludge is fed into the center of the equipment where the solids are thrown against the wall of the bowl which is rotating at high speeds, thus generating high centrifugal forces. The solids deposited against the bowl continuously move to one end of the machine by an internal screw conveyor which rotates at a different speed. The solids are then discharged into an inclined screw conveyor and dropped into a roll-off bin. The clarified liquid continuously overflows through adjustable weirs at the other end of the machine and is discharged to the WW sewer net.

All the operation phases will be fully automated and difference in the details will depend on the selected contractor. An automatic control should be installed to ensure that the centrifuge stops operations in case the slurry feed pump turns off. The unit will have integral level controls and instrumentation for proper automation through a dedicated PLC system included in the unit.

A periodical automatic high-pressure total washing cycle with water free of solids (reuse or clean water) should be performed indicatively every three to ten working days, this also depending on the characteristics of the sludge.

4.17 Gravity Thickener

A gravity thickener, installed upstream of the dewatering unit, will be used to remove part of the water and to concentrate the solids content of the sludge by gravity settling, thus reducing the volumes of sludge to be handled by downstream units. In addition, the thickener can provide storage capacity for the sludge, allowing the non-continuous operation of downstream dewatering units. This is the reason why the gravity thickener is mostly found as the first step of the solids management process. It performs the following operations:

- Increases the solid content of the sludge (i.e. increases the TSS concentration, with efficiency to be determined based on the upstream TSS content and based on HRT, by removing a portion of the liquid fraction by gravity separation).
- Sludge volume reduction as a consequence of the above process.
  The unit will be provided in one single unit. The gravity thickener should include:

- Centrifugal feed pump.
- Thickener tank.
- Moving bridge or equivalent sludge collecting mechanism.
- Thickened sludge extraction pump (progressing cavity type).
- Torque monitoring system.
- Depending on the final destination of the supernatant, a dedicated pump station could be required.
- All other electromechanical and I&C equipment (pumps, compressor, mixers, piping, valves, instruments, etc.) necessary for proper operation.

4.17.1 Process Parameters

Table 4.30 summarizes the main process parameters for the gravity thickener.

Table 4.30 Main process parameters for gravity thickener.

Parameter	Units	Value
Solid loading (minimum–maximum)	kg TSS/m^2/day	
Primary sludge		80–120
Chemical sludge (iron salts)		10–50
Chemical sludge (lime)		100–300
Biological sludge		20–40
Combined primary and biological sludge		40–60

(Continued)

Table 4.30 (Continued)

Parameter	Units	Value
Thickener overflow rate	$m^3/m^2/day$	
Primary sludge		15–30
Biological sludge		4–8
Combined primary and biological sludge		6–12
Thickened sludge dryness (minimum–maximum)	%	
Primary sludge		5–10
Chemical sludge (iron salts)		3–4
Chemical sludge (lime)		10–15
Biological sludge		2–3
Combined primary and biological sludge		4–6

4.17.2 Process Description

The sludge extracted from the primary clarifier, secondary clarifier, or chemical/physical treatments is transferred to the gravity thickener which performs the following operations:

- Gravity settling of TSS.
- Collection of concentrated sludge from the bottom hopper, by means of rotation scraper.
- Sludge extraction from the hopper.
- Clarified water discharging to WW drain.

Gravity thickening is one of the most common methods to increase the solids content of the sludge and it is accomplished in a tank similar in design to a conventional sedimentation tank. Normally, a circular tank is used and dilute sludge is fed to a center feed well. The feed sludge is allowed to settle and compact, and the thickened sludge is withdrawn from the conical hopper bottom. Conventional sludge-collecting mechanisms with deep trusses or vertical pickets stir the sludge gently, thereby opening up channels for water to escape and promoting densification. The supernatant flow that results is drawn off and returned to the WWTP equalization (if present) or gradually mixed with the influent WW fed to the plant head. The thickened sludge is pumped out from the hopper and sent to the dewatering unit.

The unit will be provided with a torque control system on the motor of the moving sludge collecting mechanism to avoid excessive torque. A level meter will trip the sludge extraction pump to protect the equipment to avoid running dry.

Further Reading

Chaubey, M. (2002a). Treatment of dairy effluent with fixed film bioreactor technology. *Indian Journal of Environmental Protection* 23 (4): 361–363.

Chaubey, M. (2002b). Computer design of wastewater treatment plant using fixed film bioreactor technology. *Journal of Industrial Pollution Control* 18 (1): 107–117.

Chaubey, M. (2005a). Water reuse is better alternative than sea water desalination. *Arab Water World Journal* XXIX (3): 22–23.

Chaubey, M. (2005b). Technological assessment of water treatment technologies. *Arab Water World Journal* XXIX (5): 34–36.

Chaubey, M. (2006a). Innovative methods of pre-treatment of reverse osmosis membranes. *Arab Water World Journal* XXX (8).

Chaubey, M. (2006b). Fouling prevention techniques of reverse osmosis systems. *Water Digest* I (1).

Chaubey, M. (2016). Assessment of aerobic biological technologies for wastewater treatment of F.M.C.G. industries. *Water Digest* 2016: 30–36.

Chaubey, M. (2018). Design considerations of wastewater treatment plants with MBBR technology. *Water Digest, January* 2018: 52–56.

Chaubey, M. and Kaushika, N.D. (2004). Field study of reverse osmosis process in wastewater treatment. *Environmental Pollution Control Journal* 7 (1): 6–11.

Culp, W. and Williams, H. (1980). *Wastewater Reuse and Recycling Technology*. Park Ridge, NJ: Noyes Data Corporation.

James M. Montgomery, Inc (1985). *Water Treatment, Principle & Design*. New York: Wiley.

Metcalf & Eddy, Inc (2004). *Wastewater Engineering: Treatment and Reuse*, 4e.

Zhongxiang, Z. and Yi, Q. (1991). Water saving and wastewater reuse and recycle in China. *Water Science Technology* 23: 2135–2140.

5

Advance Sustainable Wastewater Treatment Technologies

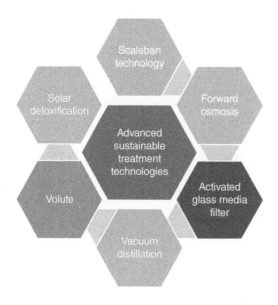

CHAPTER MENU

Managing wastewater (WW) by implementing advanced sustainable WW treatment can support achievement of the Sustainable Development Goal (SDG) target 6.3. WW should be seen as a sustainable source of water, energy, nutrients, and other recoverable byproducts, rather than as a burden. Choosing the most appropriate type of WW treatment system that can provide the most co-benefits is site specific, and countries need to build capacity to

Wastewater Treatment Technologies: Design Considerations, First Edition. Mritunjay Chaubey.
© 2021 John Wiley & Sons Ltd. Published 2021 by John Wiley & Sons Ltd.

assess this. This chapter provides brief descriptions of the latest advanced sustainable WW treatment technologies. It is based purely on the experience of the author of this book, who is involved in the piloting and customization of these technologies.

5.1 Scaleban

Scaleban is a unique and patented technology that helps industries achieve water conservation and zero liquid discharge (ZLD). It does this by integrating process effluent and reverse osmosis (RO) reject water having high total dissolved solids (TDS) with existing cooling towers in place of freshwater, by using the cooling tower as a natural evaporator without affecting the plant's performance relating to hard water scaling, corrosion, and biofouling in the cooling tower circuit. With the Scaleban system, cooling towers can be operated at higher TDS; hence effluent treatment plant (ETP)-treated water/effluent can be used as cooling tower makeup water, thus reducing raw water consumption without requiring any extra energy input for its operation.

Scaleban technology (a combination of Scaleban equipment and Scaleban specialty chemicals and filtration system) has the unique strength of preventing hard water scaling, corrosion, and biofouling in the cooling tower circuit at a TDS level of even 300 000 ppm. A schematic diagram of recycling WW with Scaleban technology is shown in Figure 5.1.

Scaleban addresses common problems associated with the use of WW in cooling towers as makeup water in the following manner:

- **Hard water scaling**: Scaleban equipment is installed at the inlet of every heat exchanger in the cooling tower circuit, which prevents hard water scaling problems in the heat transfer area.
- **Total suspended solids (TSS) removal**: An activated glass media-based side-stream filter is installed which reduces TSS load in circulation water of the cooling tower.

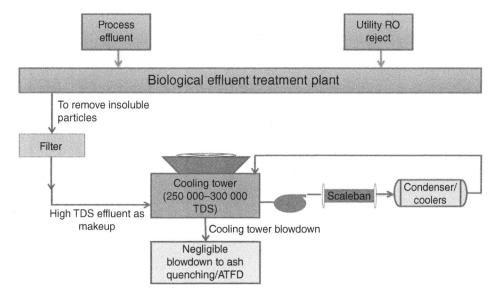

Figure 5.1 Schematic diagram for recycling WW with Scaleban technology.

- **Corrosion**: The cooling tower is operated at high hardness/TDS in alkaline zones with positive Langlier Saturation Index (LSI) values to prevent corrosion. Scaleban specialty corrosion inhibitors are dosed in the cooling tower to maintain acceptable corrosion rates.
- **Biofouling**: The operating pH range of 9–10 under high chemical oxygen demand/biological oxygen demand (COD/BOD) environments retards the growth of bacteria in recirculating water. Scaleban specialty formulated chemicals for biological control do not allow biofouling or algae formation in the cooling tower circuit.

5.1.1 Scaleban Working Principle

To understand the working process of Scaleban, let us first understand the solubility characteristics of calcium and magnesium in water. Take the example of the solubility of sugar and common salt in water. We all know that sugar and common salt dissolve in water and the rate of their solubility increases with increase in temperature. Thus we can establish that solubility increases with rise in temperature. However, in the case of calcium and magnesium the solubility shows a reverse nature with respect to temperature and pH, i.e. the solubility of calcium and magnesium in water decreases with increase in temperature or pH of water. The relationship can be seen in Figure 5.2.

The main reason for formation of scale inside any equipment is the rise in temperature of water flowing through it, since at higher temperature water has a tendency to precipitate dissolved calcium and magnesium salts. A graphical representation of the solubility of these salts in water vis-à-vis its temperature indicates that with the rise in temperature from T_1 to T_2 the solubility of these salts reduces from S_1 to S_2 (see Graph 1 in Figure 5.2). Based on scientific studies, it is known that the solubility of calcium and magnesium in water decreases with increase in pH of water. This can be seen in Graph 2 in Figure 5.2.

Similar to its tendency to precipitate out dissolved calcium and magnesium salts at higher temperature, water also tends to precipitate out dissolved calcium and magnesium salts at higher pH value, i.e. when it tends to be slightly alkaline. Graph 2 in Figure 5.2 illustrates this. With the rise in pH value from P_1 to P_2, the solubility of these salts reduces from S_1 to S_2, causing their precipitation at higher pH value.

Scaleban makes use of the galvanic principle to locally/temporarily increase the pH value of water flowing through it and precipitates out hardness-causing salts in fine particles of colloidal nature before water reaches the high-temperature zone. The galvanic principle on which Scaleban works is shown in Figure 5.3.

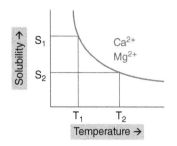

Graph 1: Relationship of temperature and solubility of Ca^{2+} and Mg^{2+} in water

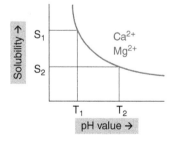

Graph 2: Relationship of pH and solubility of Ca^{2+} and Mg^{2+} in water

Figure 5.2 Solubility of Ca^{2+} and Mg^{2+} vs temperature and pH.

Galvanic principle:

"Galvanic principle says that when you immerse two electrodes of different electro-negativity in an electrolyte, electrons will start flowing from electrode of higher electro-negativity to the electrode of lower electro-negativity through the electrolyte."

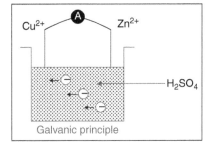

Galvanic principle

Figure 5.3 The galvanic principle on which Scaleban works.

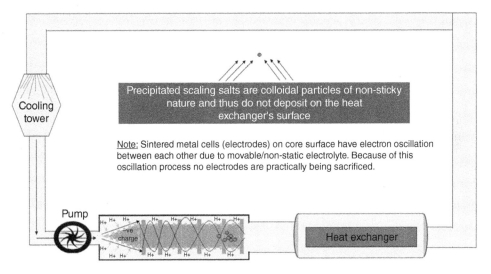

Figure 5.4 Scaleban working process in a cooling tower circuit.

Scaleban is an online pipe-shaped mechanical device which houses a very specially designed core that is sintered with a number of electronegative elements in increasing order of electronegativity in the direction of water flow. When water passes through Scaleban it acts as an electrolyte (being a self electrolyte by nature) while it comes in contact with the core placed inside Scaleban. The Scaleban working process in a cooling tower circuit is shown in Figure 5.4.

Due to galvanic action, the core gets negatively charged and attracts the lightest ions present in the fluid, i.e. hydrogen ions (H^+), toward itself. The relationship between pH of water and H^+ ion is expressed by the formula PH α $1/H^+$. As pH is inversely proportional to the concentration of H^+ ions, concentration of H^+ ions becomes comparatively lower in the fluid inside the Scaleban equipment, pH value increases, and this results in precipitates of calcium and magnesium in very fine particles having a colloidal nature, before they come into contact with heat transfer surfaces. Once the precipitation of these hardness-causing salts has already taken place, further precipitation of hardness-causing salts which are responsible for scaling is not possible due to further rise in temperature in heat exchangers/condensers. This results in the scale-free operation of condensers/heat exchangers for 20 years. These fine colloidal nature hardness-causing salts have a tendency to redissolve in water when they reach zero velocity zones in the cooling tower basin.

5.2 Forward Osmosis

Forward osmosis (FO) is a natural process and an integral part of the survival of flora and fauna on this planet. It is this process only that makes plants transport water from their root systems to their leaves and it provides the primary means of transporting water in cells across most organisms in nature. Both FO and the conventional RO process are highly selective for water molecules. The difference lies in the means by which water molecules are driven through the membrane. The FO process is governed by the difference in osmotic pressure, and the direction of water diffusion takes place from lower concentration (the feed side) to higher concentration (the draw side). RO processes, on the other hand, are governed by hydraulic pressure differences and the direction of water diffusion is from high concentration to low concentration. The FO process employs semipermeable membranes to concentrate the dissolved contaminants and separate fresh permeate from dissolved solutes. The driving force for this separation is an osmotic pressure gradient which is generated by a draw solution of high concentration to induce a net flow of water through the membrane into the draw solution, thus effectively separating the feed water from its solute. As osmosis is a natural phenomenon, it significantly requires less energy compared to the conventional RO process. The principle of FO is shown in Figure 5.5.

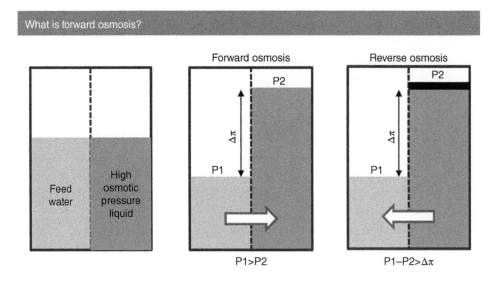

Figure 5.5 Principle of forward osmosis.

Figure 5.6 Forward osmosis plant.

FO is one of the best technologies to treat high TDS WW (5000 ppm < TDS > 100 000 ppm) and moderate COD WW (10 000 ppm < COD > 20 000 ppm). A photograph of the world's first FO technology plant installed in the chemical industries by UPL Limited at Ankleshwar is shown in Figure 5.6.

The FO plant comprises three stages, namely filtration, the FO membrane system, and the draw solution recovery system. First, the high TDS effluent stream is passed through a series of filters to remove all suspended solid from the system. The filtered stream is then fed into the RO systems (two stage or three stage) to increase its concentration. For the two-stage RO system, draw solution (brine) can be concentrated up to 12.5% TDS; for the three-stage RO system, draw solution (brine) can be concentrated up to 16.5% TDS. This concentrated stream and the draw solution are fed into the FO membrane in countercurrent mode from two different sides of the membrane. Due to osmotic pressure, water from the concentrated stream travels toward the draw solution side and is separated through the draw solution RO system. The concentrated FO reject is evaporated, crystallized, and sent for secure landfilling. A typical process flow diagram of an FO plant is shown in Figure 5.7.

5.2.1 FO Technology Benefits

- Can be used for highly saline waters which are impossible to be treated through conventional RO process.
- Concentration of TDS can be increased to the tune of 16–17%.
- Since natural osmosis process is used, power consumption is relatively less compared to other conventional processes.
- Operation and handling are much easier and reliable.

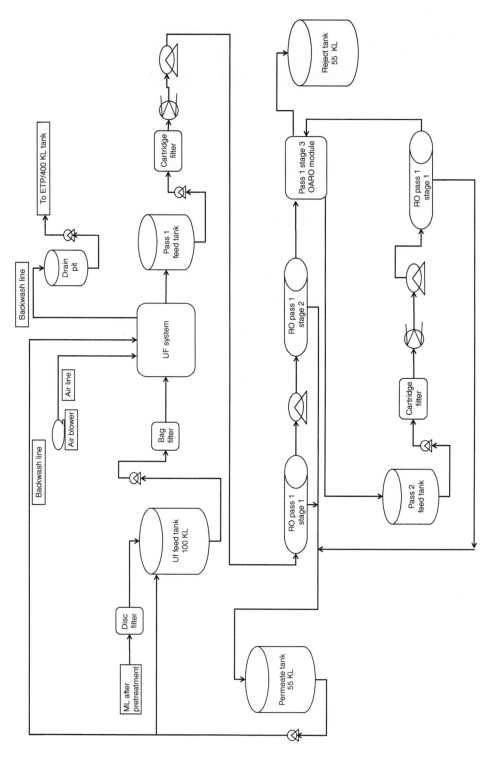

Figure 5.7 Typical process flow diagram of an FO plant. KL, kilo liter; ML, mixed liquor; OARO, osmotically assisted resverse osmosis.

5.3   Activated Glass Media Filter

Activated glass media filter (AGF) (see Figure 5.8) is a direct replacement for pressure sand filter (PSF), doubling the performance of sand filters without the need of additional investments in infrastructure. Glass media resists biofouling, biocoagulation, and transient wormhole channeling of unfiltered water and never needs to be recharged or replaced.

Figure 5.8   Activated glass media.

AGF is a highly engineered product manufactured from a specific filter media type based on alumino silicate material and is activated in a three-stage proprietary physiochemical process, processed to obtain the optimum particle size, shape, and charge. It is then exposed to a three-step activation process to increase its surface area by up to 300 times for superior mechanical and electrostatic filtration performance.

5.3.1   Process Parameters

Table 5.1 summarizes the main process parameters for the screening section. Table 5.2 summarizes the AGF product specifications. Table 5.3 summarizes the grades and arrangement from top to bottom in AGF.

Figure 5.9 shows the arrangement of AGF.

Table 5.1   Main process parameters of AGF.

Parameter	Units	Design value
Filtration velocity	$m^3/h/m^2$	1–30
Backwash velocity	$m^3/h/m^2$	<45
Maximum operating differential pressure	Bar	<0.5
Backwash duration	Minutes	3–10
Water pH limits		4–10
Water temperature limits	°C	1–100

Table 5.2 AGF product specifications.

Specification	Grade 0	Grade 1	Grade 2	Grade 3
Particle size	0.25–0.5 mm	0.4–1.0 mm	1.0–2.0 mm	2.0–4.0 mm
Undersized	<5%	<5%	<10%	<10%
Oversized	<5%	<5%	<10%	<10%
Effective size (expressed as d10)	0.3 mm	0.45 mm	1.1 mm	2.1 mm
Uniformity coefficient (d60/d10)	1.3–1.4	1.6–1.8	1.4–1.5	1.4–1.5
Aspect ratio	2–2.4	2–2.4	2–2.4	2–2.4
Organic contamination	<50 g/t	<50 g/t	<50 g/t	<50 g/t
Specific gravity (grain)	2.4 kg/l	2.4 kg/l	2.4 kg/l	2.4 kg/l

Table 5.3 Grades and arrangement from top to bottom in AGF.

Grade, size (mm)		Commercial purification	High purification	Ultra purification
Grade 0	0.25–0.50	n/a	20%	60%
Grade 1	0.4–1.0	70%	50%	20%
Grade 2	1.0–2.0	15%	15%	10%
Grade 3	2.0–4.0	15%	15%	10%

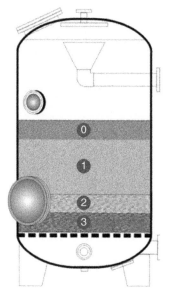

Figure 5.9 Activated glass media arrangement.

5.3.2 Process Description

WW from the upstream treatment will be fed in flow control mode (centrifugal pump equipped with variable frequency driver [VFD]) to the AGF filters. The AGF filter stage will allow further reduction of the residual TSS load present in the biological effluent.

Water to be treated will be applied to the top of the vertical bed cylindrical filter and withdrawn at the bottom. The glass media bed will be held in place with an underlying drain system at the bottom of the column. The feed system should allow homogeneous feeding of the filtering bed through the glass media and water percolation through the bed should avoid the formation of preferred filtering paths and consequent "dead zones" within the filtration bed.

Filtered water from the AGF unit will then be stored in a dedicated tank and pumped by means of centrifugal pumps (one duty, one standby) to the following ultrafiltration (UF) unit. AGF filtrate tank sizing should take into account the UF filtrate production pauses due to the backwashing and chemical cleaning cycles.

Decay of the filtration capacity of the sand bed due to its progressive saturation will be monitored by a differential pressure controller (if excess suspended matter is accumulated, the head loss through the filter increases until a maximum predetermined limit [1 bar] that will activate filter backwashing). However, automatic backwashing is normally activated once per day on a time-based logic (typically 15 minutes of backwashing).

As on option, a further control can be performed by an in-line TSS analyzer installed on the filtered water line. The sand bed will have to be backwashed when the TSS value and/or the pressure drop increases to the set high limit for the effluent quality. Backwashing will allow the expansion of the sand layer so that the solids captured during normal operation can be released and separated from the sand. Backwashing water will be sent back to the equalization tank of the wastewater treatment plant (WWTP). The water used for the backwashing will be taken from the UF filtered water storage tank.

AGF configuration will be constituted by a minimum of two filters in parallel which will be set up and provided with automatic valves in order to allow automatic switch of the operating units order within the filtration cycle: in case one AGF filter is out of operation or in backwashing, a valve will divert the total flow to the other one to ensure continuous water filtration. This is an optional case. Normally one AGF will be in operation.

The unit will have integral instrumentation for proper automation through a dedicated programmable logic controller (PLC) system, as shown in Figure 5.10.

Figure 5.10 Activated glass media filter.

5.3.3 AGF Benefits

- More than doubles the performance of an existing filtration system without the need for additional investments.
- Substantially lowers chlorine oxidation demand by up to 50%.
- Lowers backwash water demand by an average of 50%.
- Lower backwash time.
- Ninety-five percent removal of 4-μ particles with standard glass media Grade 1 and 97% of 1-μ particles with Grade 0, independently verified.
- When combined with coagulation and flocculation, nominal filtration performance is better than 0.1 μ.
- Unlike sand there is no free silica in treated water, so less fouling of membranes.
- Performance comparable to UF.
- Does not biofoul or channel.
- Directly adsorbs organics like activated carbon.
- Chlorination is not required, so no trichloroacetic acid (TCA), trihalomethane (THM), or hypobromous acid production.
- Regenerated by backwashing with water; air purge not required.
- Perfect for groundwater filtration and removal of heavy metals like iron, manganese, and arsenic.

5.4 Vacuum Distillation

ZLD has become a necessity for many of the industries as this not only eliminates the need to monitor violations but also helps in conserving almost every drop of water that is generated as waste from industrial processes.

Once the effluent that is generated is treated by biological systems, the treated water is further taken into a UF/RO system for reduction of TDS. This creates a high TDS (reject) and low TDS (permeate) water. Most of the time, the permeate is limited to 60–70%. However, some modern technologies allow 90%+ recovery. Either way the reject needs to be handled. Also, in some cases the effluent generated itself contains high TDS and the biological treatment becomes unviable or not possible.

High TDS in treated wastewater is not allowed by regulatory agencies as such water has a tendency to contaminate the receiving water and increase soil salinity, making the receiving water and land saline. Continued disposal of high TDS water on land increases the groundwater salinity level, gradually making the groundwater unsuitable for use. The maximum limit for TDS in treated wastewater is 2100 ppm, beyond which it is not even suitable for gardening.

Absolute removal of TDS is possible only by evaporation and crystallization, which is both capital expenditure (CAPEX) and operating expenses (OPEX) intensive. Hence water footprint reduction, wastewater volume reduction, recovery, and treatment, and recycling to identified reuse by adopting an integrated approach to water and wastewater management are critical. Though various evaporation and crystallization technologies and options are available for treating wastewater having high TDS they all have advantages and disadvantages and careful evaluation has to be done to meet the final objectives, which in most cases will be site and industry specific.

Some technologies use natural renewable sources of energy like wind and solar and some technologies use non-renewable sources of energy like coal, electricity, gas, and oil. The technology of choice depends on various factors, including but not limited to:

- Flow rate and volume of wastewater.
- Characteristics of wastewater.
- Space.
- Budget.
- Ease of use.
- Operating cost.
- Capital cost.
- Operational complexity.
- Manpower requirement.
- Maintenance requirement.
- Robustness and reliability.
- End user preference.

Now that waste disposal is increasingly unacceptable, disposal of parts cleaning WW is frequently done most effectively by evaporation, due to process problems inherent in membrane and chemical systems and the undesirable cost of off-site disposal. Thus, evaluation of total evaporation costs is important for assessing total cleaning cost. Consider the following evaporation disposal costs:

- Evaporator maintenance (removal of accumulated residues, cleaning of mist eliminators, heat exchangers, mid-process oil removal, and/or settled solids blowdown).
- Evaporation process upsets (due to variations in WW oil loading, solids loading, or foam formation during evaporation).
- Periodic evaporator operator training and preventive maintenance (PM) (on the evaporator itself).
- Normal operating energy cost (based on fuel consumption required). No evaporator can perform at 100% efficiency.

5.4.1 Process Upsets

Evaporation process upsets are harder to quantify; but, again, the evaporator manufacturer's estimates based on their experience are key to appraising this cost element based on your WW's estimated range of oil and solids loading. On the other hand, only actual testing of your typical waste will be appropriate in assessing the impact of foam, a process upset that can severely limit your WW throughput. This may require:

- Adjusting heat input to the system.
- Adjusting normal liquid operating level within the evaporator.
- Adding chemical antifoam chemistry periodically.

All of these obviously impact operating cost, but potentially most significant is the requirement for antifoam chemistry. Be sure your evaporator supplier provides a foam assessment and recommends a specific chemical and dosage rate to permit your evaluation of this cost element, which very much depends on the evaporator design. Training and project management WW evaporators generally take little time to start and operate. New operators generally become comfortable with very little training. However, PM requires more effort. Since the evaporator is frequently considered a "garbage can" for liquids, it can be badly abused. Without sufficient control over what is placed in it, what conditions need monitoring, and – most of all – how often residues must be removed, the evaporator can become subject to extensive downtime for correction of easily avoidable conditions. Most common among user concerns are conditions caused by over-concentrating non-water constituents in the waste. If you were evaporating only water (instead of WW) there would be little concern. But then also there would likely be no need for an evaporator. It is the presence of waste contaminants that creates WW that is unacceptable to the sewer. As with items 1 and 2 above, these training and PM costs must be discussed with the evaporator supplier to assess likely hours that will be consumed in establishing good evaporator habits. Service agreements with the evaporator manufacturer can level out fluctuations in PM costs and provide a better-controlled WW disposal system consistent with ISO 9001 and 14001.

There is an energy cost of evaporation. Technology is beautiful when it is simple! Finally, the energy cost of evaporation (frequently the only cost considered) must be estimated. But here there is room for much more precision in calculating cost. Your firm's present rate of energy consumption establishes a level of cost-per-unit-energy that, together with the evaporator's spec sheet, allows a very close estimate of operating cost. That is, of course, assuming the spec sheet "tells all.". It is possible to understate energy requirements and thus overstate efficiency. You can conquer that engineering calculation with a little help in understanding evaporator efficiencies.

Those who are contemplating the use of a WW evaporator should understand that the energy consumption required to evaporate a given volume of water can vary significantly from one evaporator design to another. It is important to remember that the efficiency of the evaporative equipment (output divided by input) will determine a significant portion of the actual operating cost.

There are several different ways that this information can be presented and applied. The most meaningful is by comparing the amount of fuel delivered to the equipment with the amount of water that is evaporated over a given time. Evaporation and condensation has been the preferred and proven method for such recovery.

Conventionally, a multiple-effect evaporator (MEE) has been the most widely used for such treatment. This happens in two stages. First, a three-effect evaporation is used to make steam from the effluent, and distillate is recovered. This will give roughly 65% recovery. In the second stage the balance (35%) is taken into an agitated thin film drier and distillate is recovered. You get dry salts/solids that can be disposed in a controlled landfill.

This system comes with several disadvantages as follows:

- A steam source is required. This not only calls for an expert team but also comes with mandated periodic inspectorate approvals.
- Efficiency of the MEE drops every year progressively and pretty quickly.
- Jet cleaning is required every month.

- Usage of fossil fuels negates the carbon credit earnings.
- It requires expert maintenance.

5.4.2 Principle of Operation

The influent after passing through a micron filter is taken into a jacketed vacuum chamber. A compressor connected to the top of the chamber sucks out the vapors. The vapor is compressed there by increasing the temperature of the vapor. This is then circulated through the jacket where the heat transfer takes place with the influent. When the vapor reaches the saturation point, this is passed through the baffles and comes out as distillate. When the influent inside the chamber loses moisture and turns into a slurry, it is drained as concentrate. The drained concentrate is allowed to still and the supernatant goes back into the influent chamber. The settled slurry will have 20% consistency and can be disposed of directly or further dried using a filter press or vacuum drier.

Vacuum distillation (see Figure 5.11) typically runs for 22 h/day and then goes into cleaning in place (CIP) mode. Here diluted neutralizer and descalents are used to flush the system. This concoction is reused several times before disposal there by optimizing chemical consumption. The complete operation is automatic and does not require any human intervention. Once the CIP is complete the vacuum distillation goes into start mode. It may take anywhere between 30 and 45 minutes to start and stabilize.

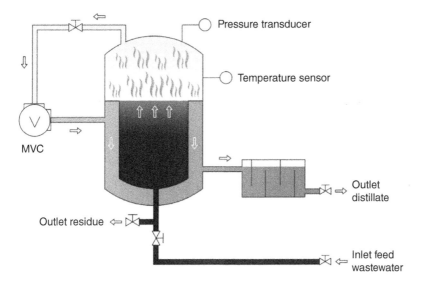

Figure 5.11 Schematic of vacuum distillation.

5.4.3 Process Description

Figure 5.12 shows a typical vacuum distillation unit. Process WW is fed by gravity through the feedstock heat exchanger and into the circulating stream. The feedstock heat exchanger

Figure 5.12 Vacuum distillation.

is used to heat the WW by transferring sensible heat from the hot distillate/condensate to the cooler feed.

- WW circulates from the heat exchanger bottom tank through the main heat exchanger tubes and back into the separation tank. Latent heat from the compressed vapor is transferred to the WW via the main heat exchanger.
- The downstream pressure is low enough to allow flashing of the circulating stream into liquid and vapor components.
- The vapor stream exits the tank at the top and flows to the vapor compressor(s).
- The vapor compressor compresses the vapor (raising the temperature and pressure), and sends the vapor to the main heat exchanger, where it transfers its latent heat to the WW in the recirculation loop.
- High-temperature condensate exits the main heat exchanger and flows to the condensate tank, where any remaining vapor is separated. The hot condensate is then pumped to the feedstock heat exchanger, where it transfers sensible heat to the incoming feed WW.
- The self-cleaning heat exchanger keeps the heat transfer surface clean and improves the energy efficiency up to 28% with the help of circulating ceramic media.
- Upon reaching steady-state at the target concentration, the concentrated WW is purged from the bottom with the help of compressed air, using the residue valve.

Table 5.4 Brief comparison of vacuum distillation and MEE technology.

Parameter	Vacuum distillation ZLD	MEE ZLD
Energy requirement	3.0 KWH for 40l	28.5 KWH for 40l
Source of energy	Electricity only	Boiler fuel and electricity
Components of ZLD system	Akvazen only	1) Boiler 2) Cooling tower 3) Evaporator
Space requirement	Single unit	Space is required for three components and piping between them
Water requirement	No additional water requirement	Additional water is required to generate steam which is used to evaporate
Compressed air requirement	Required for control purposes	Not required
Ease of installation	Less than a day required to set up	Typically takes a week to set up

- The inbuilt foam sensor detects if there is foaming inside the heat exchanger and adds defoaming chemicals automatically as and when required.
- The self-cleaning heat exchanger reduces the need for frequent CIP and increases the performance and reduces the down-time for cleaning.
- Inbuilt process control and automation measures all key process parameters such as temperature, vacuum levels, level, flow, and volume and ensures process safety, equipment safety, and operator safety.
- Real-time remote monitoring technology ensures the evaporator works at optimum level and helps in automatic data logging, monitoring, trouble shooting, etc.
  Table 5.4 gives a brief comparison of vacuum distillation and MEE technology.

5.5 Volute

Volute is a sludge dewatering equipment which removes water and moisture from sludge on a continuous basis. This multi-disc sludge dewatering press consists of two types of rings: a fixed ring and a moving ring. A screw thrusts the rings together and pressurizes the sludge. Gaps between the rings and the screw pitch are designed to gradually get narrower toward the direction of sludge cake outlet and the inner pressure of the discs increases due to the volume compression effect, which thickens and dewaters the sludge. The volute press working process is described in Section 3.5.5 and a typical volute is shown in Figure 3.26.

5.5.1 Advantages

- **Easy operation and maintenance**: Intuitively understandable operation system adopted. Monitoring of the operation settings is made very easy and 24-hour unattended operation is possible with no daily maintenance.

- **Water-saving**: Volute prevents clogging with its unique self-cleaning mechanism, removing the need for huge amounts of water for clogging prevention.
- **Power saving**: The screw which is the main component of volute rotates very slowly at a rate of 8–10 rpm, so that it consumes very low power and is thus economical. This equipment consumes the lowest energy of all dewatering equipment.
- **Low noise/low vibration**: Because volute has no rotating body with high speed, there is no concern about noise and vibration. A comfortable work environment can be secured. During standard operation, noise level is as low as approx. 65 dB.
- **High resistance to oily sludge**: The self-cleaning mechanism means volute is ideal to dewater oily sludge, which easily causes clogging and is difficult to treat with other types of dewatering equipment.
- **Small footprint**: Volute can be installed in places where placement would not be possible with other technologies. This makes volute suitable for customers who are considering the replacement of existing dewatering equipment.
- **Wide range of applications**: Applications include municipal sewage sludge, sludge generated from various industrial ETPs, primary/chemical sludge, biological sludge, oily sludge, and dissolved air flotation (DAF) sludge.

5.6 Solar Detoxification

The removal of toxic elements and organic compounds from WW using catalyst in the presence of ultraviolet (UV) radiation may be referred to as solar detoxification. Recently, it has been demonstrated that solar detoxification has great potential for the elimination of toxic elements, organic compounds, and biological contamination in WW [1–7]. Solar detoxification is a process of treatment of WW in which titanium dioxide (TiO_2) is exposed to the sun, the catalyst absorbs the high-energy photons in light from the UV portion of the solar spectrum, and reactive chemicals known as hydroxyl radicals are formed. These radicals are powerful oxidizers and disinfectants. A concentration of 0.01% TiO_2 is most effective in killing bacteria (*Serratia marcescense*, *Escherichia coli*, and *Streptococcus aureus*).

Possible reaction pathways for solar detoxification are as follows: when a photon with an energy equal to or more than the band-gap of TiO_2 is absorbed on its surface, it causes excitation of an electron from the valence band (vb) to the conduction band (cd), forming a "positive hole" in the vb. Both the hole and the electron are highly energetic and hence highly reactive. The excited electron and the positive hole either recombine and release heat or migrate to the surface, where they can react with adsorbed molecules and cause either a reduction or oxidation of the adsorbate. Since recombination in the bulk or on the surface is the most common reaction, the quantum yields (molecules reacted/photons absorbed) of most photocatalytic reactions are low. The "positive holes" cause oxidation of the surface absorbed species, while the electrons cause reduction. Both reactions must take place in order to maintain electroneutrality. Thus, if the objective is the oxidation of organics, the electrons must be consumed in a reduction reaction such as absorption by oxygen molecules to form superoxide, in order to keep the holes available for oxidation. On the other hand, if the objective is the reduction and recovery of metals, all other reducible species such as oxygen must be eliminated or kept away. The following equations [8, 9] describe the oxidation and reduction reactions:

$$TiO_2 \xrightarrow{hv} TiO_2\left(e_{cb} + hole_{vb}\right)$$

$$TiO_2\left(e_{cb} + hole_{vb}\right) \xrightarrow{recomb.} TiO_2 + heat$$

$$H_2O \rightarrow OH + H$$

$$hole_{vb} + OH_{ads} \rightarrow OH$$

$$O_2 + e_{cb} \rightarrow O_2$$

$$2O_2 + 2H_{aq} \rightarrow H_2O_2 + O_2$$

Oxidation of organics:

$$OH + Organics + O_2 \rightarrow Products\left(CO_2, H_2O, etc\right).$$

Reduction of metals:

$$ne^{-cb} + M^{n+} \rightarrow M^o$$

5.6.1 Solar Radiant Energy

The Earth receives 1.7×10^{14} kW of energy from the sun [10] – this means 1.5×10^{18} kWh/year, or approximately 28 000 times the world consumption for one year. The extraterrestrial radiation supplied by the sun that reaches the Earth is summarized in Table 5.5.

Table 5.5 Extraterrestrial radiation supplied by the sun.

Range	Wavelength (nm)	Energy (w/m^2)	Fraction of energy
Ultraviolet	<400	95	7%
Visible	400–800	640	48%
Infrared	>800	618	45%

Source: [9].

5.6.1.1 Ultraviolet Radiation

Short wavelength ultraviolet (UV) radiation exhibits more quantum properties than its visible or infrared counterparts. UV light is arbitrarily broken down into three bands, according to its anecdotal effects. UV-A (315–400 nm), which is the least harmful type of UV light, because it has the least energy, is often called black light and is used for its relative harmlessness and its ability to cause fluorescent materials to emit visible light – thus appearing to glow in the dark. UV-B (280–315 nm) is typically the most destructive form of UV light, because it has enough energy to damage biological tissues, yet not quite enough to be completely absorbed by the atmosphere. UV-B is known to cause skin cancer. Since the atmosphere blocks most of the extraterrestrial UV-B light, a small change in the ozone layer could dramatically increase the danger of skin cancer. UV-C (100–280 nm) is almost completely absorbed in air within a few hundred meters. When UV-C photons collide with oxygen atoms, the energy exchange causes the formation of ozone. UV-C is never observed in nature, however, since it is absorbed so quickly. Germicidal UV-C lamps are often used to purify water because of their ability to kill bacteria.

5.6.1.2 Visible Radiation

Visible radiation is concerned with the radiation perceived by the human eye. The lumen (lm) is the photometric equivalent of the watt, weighted to match the eye response of the standard observer. Yellowish-green light receives the greatest weight because it stimulates the eye more than blue or red light of equal radiometric power (1 W at 555 nm = 683 lm). To put this into perspective, the human eye can detect a flux of about 10 photons/second at 555 nm; this corresponds to a radiant power of 3.58×10^{-18} W. Similarly, the eye can detect a minimum flux of 214 and 126 photons/second at 450 nm and 650 nm, respectively.

5.6.1.3 Infrared Radiation

Infrared radiation contains the least amount of energy per photon of any other band and is unique in that it has primarily wave properties. This can make it much more difficult to manipulate than UV and visible light. Infrared is more difficult to focus with lenses, refract with lenses, diffracts more, and is difficult to diffuse. Since infrared light is a form of heat, far infrared detectors are sensitive to environmental changes – such as a person moving in the field of view. Night vision equipment takes advantage of this effect, amplifying infrared to distinguish people and machinery that are concealed in the darkness.

5.6.2 Photodegradation Principles

A brief outline of photodegradation principles is presented here. In order for photodegradation to take place, photons of light must be absorbed. The energy of a photon is given by:

$$E = \frac{hc}{N}$$

where

E = Energy of photon in joules
h = Planck constant, 6.626×10^{-34} joule-sec
C = Speed of light, 3×10^{8} m/s.
N = Wavelength in m

For a molecule's bond to be broken, E must be greater than the energy of that bond. According to the Planck equation, the radiation able to produce the band gap must be of a wavelength (N) equal or lower than that calculated by the equation:

$$N = \frac{hc}{E_G} = \frac{1240.7465}{E_G} \left(\text{Take 1ev} = 1.6021 \times 10^{-19} \text{joule} \right)$$

where

E_G = Semiconductor energy band-gap in ev

h = Planck constant, 6.626×10^{-34} joules-sec

C = Speed of light, 3×10^8 m/s.

N = Wavelength in nm

Wavelengths corresponding to various energy band gaps of semiconductors are shown in Table 5.6.

Whenever different semiconductor materials have been tested under comparable conditions for the degradation of the same compounds, TiO_2 has generally been demonstrated to be the most active [11]. Only ZnO is as active as TiO_2. However, ZnO dissolves in acidic solutions which make it inappropriate for technical applications [2]. TiO_2, on the other hand, is insoluble, has strong resistance to chemicals and photocorrosion, and its safety and low cost limit the choice of convenient alternatives. Furthermore, TiO_2 is of special interest since it can use natural solar UV radiation. Other semiconductors like cadmium sulfide (CdS) absorb larger fractions of the solar spectrum, but these photocatalysts are degraded during the repeated catalyst cycles.

The most important features of TiO_2 making it applicable to the treatment of contaminated effluents are as follows:

- The process takes place at ambient temperature.
- Oxidation of the substances into CO_2 is complete.
- Oxygen necessary for the reaction is obtained from the atmosphere.
- The catalyst is cheap and innocuous, and can be reused.
- The catalyst can be attached to different types of inert matrices.

Table 5.6 Calculated wavelength corresponding to various energy band gaps of semiconductors.

Semiconductor material	Energy band gap (ev)	Calculated wavelength (nm)
$BaTiO_3$	3.3	375
CdO	2.1	591
CdS	2.5	496
CdSe	1.7	730
Fe_2O_3	2.2	564
GaAs	1.4	886
GaP	2.3	539
SnO_2	3.9	318
$SrTiO_3$	3.4	365
TiO_2	3.2	388
WO_3	2.8	443
ZnO	3.2	388
ZnS	3.7	335

- The catalyst remains unaltered during the process.
- Photon energy is not stored in the final products.

5.6.3 Oxidation Potential of Various Oxidizing Reagents

Oxidation potential is a power to oxidize the organic impurities present in WW. The higher oxidation potential of oxidizing reagent indicates higher power to reduce BOD and COD from WW. An investigation has been performed in the laboratory to discover the oxidation potentials of various oxidizing reagents and the results are shown in Table 5.7. From the results, it may be noted that, except for fluorine, the hydroxide radical (OH) obtained during the solar detoxification process has the highest oxidation potential of all other oxidizing reagents.

Table 5.7 Oxidation potentials of common substances for pollution abatement.

Oxidizing reagent	Oxidation potential (V)
Fluorine	3.07
Hydroxide radical (OH)	2.90
Ozone	2.05
Hydrogen peroxide	1.80
Chlorine dioxide	1.60
Chlorine gas	1.40
Oxygen	1.30
Hypochlorite	1.00
Iodine	0.50

5.6.4 Design Methodology of Solar Detoxification System

Solar detoxification systems for WW treatment have been developed to an extent that they can be commercialized even while research continues for improving the process further. The procedure for designing such a system is complex and requires the selection of pretreatment, design of a solar detoxification reactor, design of a catalyst agitating system, and determination of flow rate, pressure drop, daily catalyst requirement, pH control, and catalyst reuse systems.

5.6.4.1 Design of Solar Detoxification Reactor

Residence time is the most influencing parameter in the design and sizing of a solar detoxification reactor. Results of an experiment indicate that after 50 minutes of residence time, reduction of BOD was 100%, reduction of COD was 86%, reduction of iron was 82%, and reduction of *E. coli* bacteria was 100%. Hence the optimum residence time may be taken as 50 minutes for design of a solar detoxification reactor at tertiary stage of treatment.

Calculations for the volume of solar detoxification reactor are as follows:

$$V = [Q * t] * 1.1 (10\% \text{ extra for free board})$$
$$V = \frac{1.1 * Q * t}{60}$$

where:

V = Volume of solar detoxification reactor in m^3
Q = Flow of WW in m^3/h.
t = Residence time in minutes.

5.6.4.2 Design of Catalyst Agitating System

Catalyst is used in a slurry configuration, which settles to the bottom of the reactor if it is not continuously mixed. Air blowers can be used as energy-efficient catalyst agitating system.

The mixing requirement of air blowers is 20 m^3/m^3 of volume of reactor. Flow of air blower is 20 times the volume of solar detoxification reactor.

5.6.4.3 Daily Catalyst Requirement

Titanium dioxide (TiO$_2$) used as catalyst plays a very important role in the solar detoxification of organic, biological, and toxic elements. The major properties of TiO$_2$ used in this experiment are summarized in Table 5.8.

Total catalyst required per day per unit flow of wastewater is a very important parameter in the process economics. Catalysts may become less effective by contaminants in contact or may be washed away in the discharge effluent, reducing the catalyst's life and requiring additional catalyst for every day. Total catalyst requirement can be estimated by conducting tests with the actual contaminated water. In the laboratory, tests conducted on dairy effluent and domestic effluents from the automobile industry found that approximately 15% of the catalyst was lost per day. Hence it may be recommended to add 15% catalyst daily in ETPs operating for continuous 24 hours.

Calculations for daily catalyst requirement are as follows:

$$W_c = \frac{15 (Q * T * C_c * 1000)}{100 X 10^6}$$

Table 5.8 Major properties of titanium dioxide.

Properties	Value
Structure	Rutile
Surface area	6.5 m^2/g
Size	Length: 10.6 μm
	Diameter: 4.7 μm
Impurities	0.73% (aluminum)
	1.14% (chlorine)

$$W_c = 1.5Q * T * Cc * 10^{-4}$$

where

W_c = Weight of daily catalyst required in kg
Q = Flow of effluent in m^3/h
T = Total operating time in hour
C_c = Rate of catalyst concentration in ppm.

5.6.5 System Configuration of Solar Detoxification Process

The configuration of solar detoxification systems, used for removal of toxic elements, organic compounds, and biological contamination at tertiary stage of the treatment sequence of WW for zero effluent discharge is shown in Figure 5.13. It consists of a shallow solar pond reactor made of reinforced cement concrete (RCC) material for holding the effluent. Since industries already use holding ponds for microbiological treatment of WW, the same shallow ponds can be modified as reactors. The reactor should be fitted with a diffused aerator for catalyst agitation. The solar detoxification reactor can be operated in a slurry configuration. The catalyst TiO$_2$ at concentration 0.1% should be mixed with effluent in the reactor. The other important component of this system is UV radiation. During day-time natural available solar radiation can be utilized for these purposes, and during night or rainy days UV sterilizer can be used. After mixing the effluent with catalyst, and allowing for a residence time of 50 minutes, the effluent should be passed through a specially designed high-rate tube settler for clarification of catalyst. The settled catalyst should be recycled to reuse in the solar detoxification reactor.

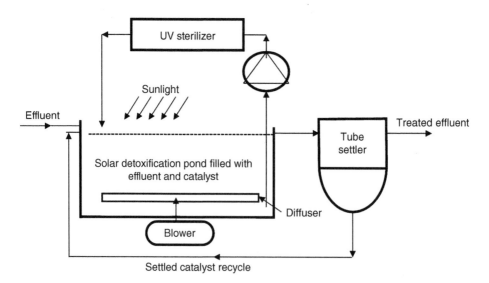

Figure 5.13 System configuration of solar detoxification process.

5.7   Sustainable Wastewater Treatment

For sustainable WW treatment of a complex WW inside a larger industrial plant, effluent characterization and stream segregation play a vital role. The WW from the manufacturing process is segregated into three streams for easier and more efficient ways of achieving sustainable treatment. The three streams are:

- Green stream.
- Yellow stream.
- Red stream.

In the green stream, effluents have a TDS below 5000 ppm and a COD less than 10 000 ppm. After this effluent is segregated it is sent for biological treatment, using either activated sludge process (ASP) technology or moving bed biofilm reactor (MBBR) technology for appropriate treatment. The treated effluent is sent to the Scaleban, where it can be utilized in the cooling towers.

In the yellow stream, effluents have a TDS below 100 000 ppm and higher than 5000 ppm, whereas the COD is below 20 000 ppm. This effluent is sent to either FO or the OH radical where advanced membrane or advanced oxidation technology is used. Usually, 10% of the effluent generated is rejected and is sent to the Scaleban, where it can also be utilized in the cooling towers, whereas the rest (90%) that is generated is clean water which can be used for other purposes at the plant or can be moved for water recycling or reuse.

In the red stream, high polluting, high COD, and high TDS effluents are generated, hence nearly 10% of total effluent. This effluent is channelized for MEE and sent for recycling or reuse.

ETP sludge and the solid waste produced throughout this process are sent to secured landfills, thus making this process sustainable for use. A sustainable treatment scheme inside a large complex is shown in Figure 5.14.

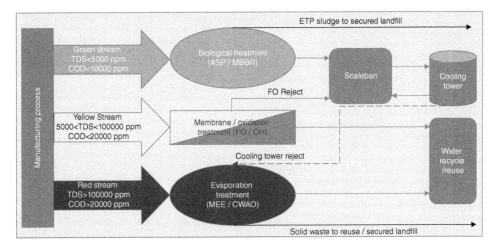

Figure 5.14   Sustainable WW treatment scheme inside a large complex.

This form of treatment is very relevant since the effluents are segregated into streams, which leads to easier, efficient, and sustainable treatment of effluents. Usually, 60% is segregated into the green stream, 30% into the yellow stream, and 10% into the red stream, but this can vary from one industry to another.

The main logic of this form of treatment is that even though the effluents in the red stream are much less in amounts, they are the most polluting and if they are not segregated they mix with the other effluents, causing extreme pollution throughout the system, which is not favorable. Hence, segregation is of utmost importance and should always be considered when treating effluents.

The application of relevant technologies is also as important as segregation. As mentioned above, treatments like ASP and MBBR in the green stream, FO and OH radicals in the yellow stream, and MEE in the red stream lead to desirable outcomes. Thus, this form of technology should be taken into consideration by companies as it is not only efficient but also sustainable in nature.

References

1 Ollis, D.F. (1994). *Photoreactors for Purification and Decontamination of Air*. Raleigh, NC: Department of Chemical Engineering, North Carolina State University.

2 Goswami, D.Y. (1995). Engineering of Solar Photocatalytic Detoxification and Disinfection Process. In: *Advances in Solar Energy*, vol. 10 (ed. K.W. Boer), 165–210. Boulder, CO: American Solar Energy Society.

3 Poulions, I., Makri, D., and Prohaska, X. (1999). Photocatalytic Treatment of Olive Milling Wastewater. *Global NEST International Journal* 1 (1): 55–62.

4 Lianfeng, Z., Tatsuo, K., Noriaki, S., and Atsushi, T. (2001). Photocatalytic Degradation of Organic Compounds in Aqueous Solution by a TiO_2 Coated RBC Using Solar Light. *Solar Energy* 70 (4): 331–337.

5 Habibi, M.H., Tangestaninejad, S., and Yadollahi, B. (2001). Detoxification of Water Containing Para Methyl Thiophenol with Photo Catalytic Oxygenation on Titanium Dioxide. *Journal of Industrial Pollution Control* 17 (1): 67–73.

6 Blake, D.M., Link, H.F., and Eber, K. (1992). Solar Photocatalytic Detoxification of water. In: *Advances in Solar Energy*, vol. 7 (ed. K.W. Boer), 167–210. Boulder, CO: American Solar Energy Society.

7 Legrini, O., Oliveros, E., and Braun, A.M. (1993). Photochemical Processes for Water Treatment. *Chemical Reviews* 93: 671–698.

8 Bahnemmann, D., Bockelmann, D., and Goslich, R. (1991). Mechanistic Studies of Water Detoxification in Iluminated TiO_2 Suspensions. *Solar Energy Materials* 24: 564–583.

9 Webb, J.D., Blake, D.M., Turchi, C., and Magrini, K. (1991). Kinetic and Mechanistic Overview of TiO_2-Photocatalyzed Oxidation Reactions in Aqueous Solution. *Solar Energy Materials* 24: 584–593.

10 Hulstrom, R., Bird, R., and Riordan, C. (1985). Spectral Solar Irradiance Data Sets for Selected Terrestial Conditions. *Solar Cells* 15: 365–391.

11 Nogueira, P.F.R. and Jardim, W.F. (1996). TiO_2 Fixed Bed Reactor for Water Decontamination Using Solar Light. *Solar Energy* 56: 471–475.

Further Reading

Chaubey, M. (2002). Treatment of industrial wastewater with solar detoxification technology. *Environmental Pollution Control Journal* 5 (3): 36–39.

Chaubey, M. (2003). Solar powered biological fouling prevention techniques of reverse osmosis membranes. *Environmental Pollution Control Journal* 6 (6): 51–54.

Chaubey, M. (2016). Assessment of aerobic biological technologies for wastewater treatment of F.M.C.G. industries. *Water Digest* 2016: 30–36.

Chaubey, M. (2019). Best practices & design considerations for wastewater treatment with MBBR technology. *Official Journal of Indian Chemical Council*: 16–19.

Chaubey, M. and Kaushika, N.D. (2003). Performance analysis of fixed film bioreactor plants for wastewater treatment. *Journal of Industrial Pollution Control* 19 (2): 203–213.

Kaushika, N.D. and Chaubey, M. (2004). Laboratory investigations of photocatalytic detoxification for the prevention of biological fouling in reverse osmosis membrane. *International Journal, Research Journal of Chemistry and Environment* 8 (2): 15–20.

6

Zero Liquid Discharge

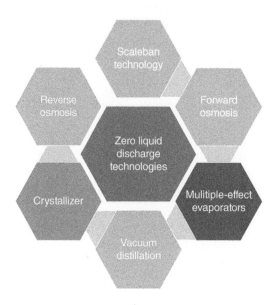

CHAPTER MENU

Zero liquid discharge (ZLD) is a wastewater (WW) management system that ensures that there will be no discharge of industrial WW into the environment. It is achieved by treating WW through recycling and then recovery and reuse for industrial purpose. Hence ZLD is a closed loop cycle with no discharge.

Water pollution by inappropriate management of industrial WW is one of the major environmental problems encountered globally. It is extremely necessary to treat as well as recycle industrial WW to reduce water scarcity and decrease the burden on natural resources. Industrial ZLD has received steadily increasing attention across the world due to

Wastewater Treatment Technologies: Design Considerations, First Edition. Mritunjay Chaubey.
© 2021 John Wiley & Sons Ltd. Published 2021 by John Wiley & Sons Ltd.

stringent regulations and water scarcity. To achieve industrial ZLD, innovative and sustainable approaches are required to select the right technology. In this chapter we discuss an innovative and sustainable approach toward industrial ZLD. This approach will help industry to address the issue of minimizing waste as well as help in adoption of cleaner production technologies. Before considering ZLD, understand your treatment goals, economics, and regulatory requirements. For example, concentrating WW to a lower volume brine that can be sent for disposal may be more cost effective than producing ZLD solids.

6.1 ZLD Technologies

There are several methods available at tertiary stage for the treatment of wastewater (WW) to achieve zero liquid discharge (ZLD). Some of the prominent ZLD technologies at tertiary stage for the treatment of WW are described below:

- Reverse osmosis (RO).
- Scaleban.
- Forward osmosis (FO).
- Vacuum distillation.
- Multiple effect evaporator (MEE) cum crystallizer.

6.1.1 Reverse Osmosis

An RO unit which is part of the tertiary treatment of the wastewater treatment plant (WWTP) is used to reduce dissolve salts from treated WW. The RO unit may include a single-stage RO system, or a double-stage system, consisting of two RO subunits that perform the following operations in series:

- First-stage RO for the treatment of wastewater coming from upstream treatments.
- Second-stage RO for the treatment of brine produced in the first RO stage.

The latter option should be considered if there is the need to maximize water reuse and a further concentration of the brine generated by the first RO stage is required in order to minimize the brine stream to be disposed of.

The RO unit will be provided in one single line but with a sparing philosophy considered to have, for each of the most critical rotating equipment, a $1 + 1$ philosophy, with an installed spare that can be automatically switched into operation in case of failure of the main one, in order to ensure the continuity of the operation of the WWTP. The RO unit should include:

- One cartridge filter 5 µm.
- High-pressure pumping unit.
- First-stage RO treatment system.
- Second-stage RO treatment system (optional).
- Antiscalant dosage system.
- Sodium metabisulfite dosage system.
- Acid dosing system (hydrochloric/sulfuric acid).
- Permeate/CIP buffer tank.

- Permeate/CIP pumping system.
- All the electromechanical and instrumentation and control (I&C) equipment (pumps, instruments, etc.) necessary for proper operation.

6.1.1.1 Power Consumption
Power consumption is in the range of 2.5–5 kWh/m^3 of product water produced.

6.1.1.2 Dissolved Solids Removal
Normally, brackish water (BW) RO membranes are used to reduce dissolved salts in WW. The RO system is generally used to reduce dissolved salts up to 10 000 ppm.

6.1.1.3 Capital Cost
In general, total capital cost is in the range of INR 50–100/m^3 of product water produced.

6.1.2 Scaleban

Scaleban is a unique and patented technology which helps industries in achieving water conservation and ZLD by integrating process effluent and RO reject water having high total dissolved solids (TDS) with existing cooling towers in place of freshwater by using cooling tower as a natural evaporator that too without affecting plant's performance related to hard water scaling, corrosion, and biofouling in the cooling tower circuit. With application of Scaleban system, cooling towers can be operated at higher TDS hence effluent treatment plant (ETP) treated water/effluent can be used as cooling tower makeup water and thus reducing raw water consumption without requiring any extra energy input for its operation.

6.1.2.1 Power Consumption
Power consumption is in the range of 0.5–2 kWh/m^3 of product water produced.

6.1.2.2 Dissolved Solids Removal
The Scaleban system is generally used to reduce dissolved salts up to 200 000 ppm.

6.1.2.3 Capital Cost
In general, total capital cost is in the range of INR 150–200/m^3 of product water produced.

6.1.3 Forward Osmosis

FO is a natural process and an integral part of the survival of flora and fauna on this planet. It is this process only that makes plants transport water from their root systems to their leaves and it provides the primary means of transporting water in cells across most organisms in nature. Both FO and conventional RO processes are highly selective for water molecules. The difference lies in the means by which water molecules are driven through the membrane. The FO process is governed by a difference in osmotic pressure and the direction of water diffusion takes place from lower concentration (the feed side) to higher concentration (the draw side). RO processes, on the other hand, are governed by hydraulic pressure differences and the direction of water diffusion is from high concentration to low concentration. The FO process employs semipermeable membranes to concentrate the dissolved contaminants and separate fresh permeate from dissolved solutes. The driving force for this separation is an osmotic pressure gradient which is generated by a draw solution of high concentration to induce a net flow of water through the membrane into the draw solution, thus effectively separating the feed water from its solute. As osmosis is a natural phenomenon, it significantly requires less energy compared to the conventional RO process.

6.1.3.1 Power Consumption

Power consumption is in the range of $7.5-10\,kWh/m^3$ of product water produced.

6.1.3.2 Dissolved Solids Removal

The FO system is generally used to reduce dissolved salts from 40 000 to 200 000 ppm.

6.1.3.3 Capital Cost

In general, total capital cost is in the range of INR $300-400/m^3$ of product water produced.

6.1.4 Vacuum Distillation

Vacuum distillation is the process of lowering the pressure in the column above the solvent to less than the vapor pressure of the mixture, creating a vacuum, and causing the elements with lower vapor pressures to evaporate off.

The influent after passing through a micron filter is taken into a jacketed vacuum chamber. A compressor connected to the top of the chamber sucks out the vapors. The vapor is compressed there by increasing the temperature of the vapor. This is then circulated through the jacket where the heat transfer takes place with the influent. When the vapor reaches saturation point, it is passed through the baffles and comes out as distillate. When the influent inside the chamber loses moisture and turns into a slurry, it is drained as concentrate. The drained concentrate is allowed to still and the supernatant goes back into the influent chamber. The settled slurry will have 20% consistency and can be disposed of directly or further dried using a filter press or a vacuum drier.

6.1.4.1 Power Consumption

Power consumption is in the range of $15-22.5\,kWh/m^3$ of product water produced.

6.1.4.2 Dissolved Solids Removal

The vacuum distillation system is generally used to reduce dissolved salts from 30 000 to 50 000 ppm.

6.1.4.3 Capital Cost

In general, total capital cost is in the range of INR $700-1500/m^3$ of product water produced.

6.1.5 Multiple-Effect Evaporator Cum Crystallizer

Evaporators are a kind of heat transfer equipment where the transfer mechanism is controlled by natural convection or forced convection. A solution containing a desired product is fed into the evaporator and it is heated by a heat source like steam. Because of the applied heat, the water in the solution is converted into vapor and is condensed, while the concentrated solution is either removed or fed into a second evaporator for further concentration. If a single evaporator is used for the concentration of any solution, it is called a single-effect evaporator system, and if more than one evaporator is used for the concentration of any solution, it is called a multiple-effect evaporator (MEE) system. In an MEE the vapor from one evaporator is fed into the steam chest (calandria) of the other evaporator. In such a system, heat from the original steam fed into the system is reused in successive effects.

Evaporation is a process of removing water or other liquids from a solution and thereby concentrating it. The time required for concentrating a solution can be shortened by exposing the solution to a greater surface area, which in turn would result in a longer residence time, or by heating the solution to a higher temperature. But exposing the solution to higher temperatures and increasing the residence time results in the thermal degradation of many solutions, so in order to minimize this, the temperature as well as the residence time has to be minimized. This need has resulted in the development of many different types of evaporators.

6.1.5.1 Types of MEE

Evaporators are broadly classified into four different categories:

1) Evaporators in which heating medium is separated from the evaporating liquid by tubular heating surfaces.
2) Evaporators in which heating medium is confined by coils, jackets, double walls, etc.
3) Evaporators in which heating medium is brought into direct contact with the evaporating fluid.
4) Evaporators in which heating is done with solar radiation.

Of these evaporator designs, evaporators with tubular heating surfaces are the most common of the different evaporator designs. In these evaporators, the circulation of liquid past the heating surfaces is induced either by natural circulation (boiling) or by forced circulation (mechanical methods).

Either evaporators may be operated as once-through units or the solution which has to be concentrated may be recirculated through the heating element again and again. In a once-through evaporator, the feed passes through the heating element only once; it is heated, which results in vapor formation, and then it leaves the evaporator as thick liquor. This results in a limited ratio of evaporation to feed. Such evaporators are especially useful for heat-sensitive materials.

In circulation evaporators, a pool of liquid is held inside the evaporator. The feed that is fed to the evaporator mixes with this already held pool of liquid and then it is made to pass through the heating tubes. The different types of evaporators are explained below.

6.1.5.1.1 Horizontal Tube Evaporators

Horizontal tube evaporators were the first kind of evaporators that were developed and that came into application. They have the simplest design of all evaporators. Such an evaporator has a shell and a horizontal tube such that the tube has the heating fluid and the shell has the solution that has to be evaporated. It has a very low initial investment and is suitable for fluids that have low viscosity and which do not cause scaling. The use of this kind of evaporator in the present day is minimal and limited to only preparation of boiler feedwater.

6.1.5.1.2 Horizontal Spray Film Evaporators

The horizontal spray film evaporator is a modification of the horizontal tube evaporator. It is a kind of horizontal falling film evaporator, and in these evaporators the liquid is distributed by a spray system. This sprayed liquid falls from one tube to another tube by gravity. In such evaporators, the distribution of fluid is easily accomplished and the precise leveling of fluid is not required.

6.1.5.1.3 Short Tube Vertical Evaporators

Short tube vertical evaporators were developed after the horizontal tube evaporators and they were the first evaporators that came to be widely used. These evaporators consist of tubes that are 4–10 ft long and 2–3 in. in diameter. These tubes are enclosed inside a cylindrical shell. In the center a downcomer is present. The liquid is circulated in the evaporator by boiling. Downcomers are required to permit the flow of liquid from the top tubesheet to the bottom tubesheet.

6.1.5.1.4 Basket-Type Evaporators

Basket-type evaporators have construction like short tube vertical evaporators. The only difference between the two is that basket-type evaporators have an annular downcomer. This makes the arrangement more economical. These evaporators have an easily installed deflector which helps in reducing entrainment.

6.1.5.1.5 Inclined Tube Evaporators

Inclined tube evaporators have tubes that are inclined at an angle of 30–45° from the horizontal.

6.1.5.1.6 Long Tube Vertical Evaporators

The long tube evaporator system is seen in more evaporators because it is more versatile and economical. This kind of evaporator has tubes that are 1–2 in. in diameter and 12–30 ft long. When long tube evaporators are used as once-through evaporators, no liquid level is maintained in the vapor body and the liquor has a residence time of only a few seconds. When such evaporators are used as recirculation-type evaporators, a particular level must be maintained in the vapor body and a deflector plate is provided. The liquor temperature in the tube is not uniform and is difficult to predict. Because of the length of the tubes, the effect of hydrostatic head is very pronounced.

6.1.5.1.7 Rising Film Evaporators

The working principle behind rising film evaporators is that the vapor traveling faster than the liquid flows in the core of the tube, causing the liquid to rise up the tube in the form of a film. When such a flow of the liquid film occurs, the liquid film is highly turbulent. Since in such evaporators the residence time is also low, they can be used for heat-sensitive substances too.

6.1.5.1.8 Falling Film Evaporators

In a falling film evaporator, the liquid is fed in at the top of long tubes and allowed to fall down under the effect of gravity as films. Heating media are present inside the tubes. The process of evaporation occurs on the surface of the highly turbulent films. In such an arrangement, vapor and liquid are usually separated at the bottom of the tubes. In some cases, vapor can flow up the tubes, in a direction opposite to the flow of liquor. The main application of falling film evaporators is for heat-sensitive substances, since the residence time in falling film evaporators is less. It is also useful in the case of fouling fluids, as evaporation takes place at the surface of the film and therefore any salt that deposits as a result of vaporization can be easily removed. These kinds of evaporators are suitable for handling viscous fluids, since they can easily flow under the effect of gravity. The main problem associated with falling film evaporators is that the fluid which has to be concentrated has to be equally distributed to all the tubes, i.e. all the tubes should be wetted uniformly.

6.1.5.1.9 Rising Falling Film Evaporators
When both rising film evaporator arrangement and falling film evaporator arrangement is combined in the same unit, it is called a rising falling film evaporator. Such evaporators have low residence time and high heat transfer rates.

6.1.5.1.10 Forced Circulation Evaporators
A forced circulation evaporator is used in cases where boiling of the product on the heating surface is to be avoided because of the fouling characteristics of the liquid. In order to achieve this, the velocity of the liquid in the tubes should be high and so high-capacity pumps are required.

Force is used to drive the liquid through the evaporator tubes, thus producing high tube velocities. A high-efficiency circulating pump, designed for large volume and sufficient head, is used to supply the force. Proper design results in controlled temperature rise, controlled temperature difference, and tube velocities that give optimum heat transfer. Forced circulation evaporators are recommended for viscous, scaling, and salting liquids. As the liquid is only heated in the steam chest with flashing taking place in the separator and no boiling taking place within the tubes, fouling on hot tube walls is reduced.

6.1.5.1.11 Plate Evaporators
Plate evaporators are constructed of flat plates or corrugated plates. One of the reasons for using plates is that scales will flake off the plates more readily than they do from curved surfaces. In some flat evaporators, plate surfaces are used such that alternately one side can be used as the steam side and liquor side, so that when a side is used as the liquor side and scales are deposited on the surface, it can then be used as the steam side in order to dissolve those scales. The benefits of plate evaporators are as follows:

- **Improved product quality**: The short heat contact period resulting from single pass operation and low liquid holdup eliminates product deterioration even when highly heat-sensitive liquids are involved. The resulting concentrate is of the highest quality possible.
- **Low liquid holdup**: Very little product is actually in the plate evaporator at any time. This permits rapid startup and shutdown with minimal waste. Small quantities of product may be processed economically.
- **Easy cleaning**: All stainless-steel surfaces within the plate packs are fully accessible for easy cleaning. Normal cleaning in place (CIP) is done with low consumption of cleaning chemicals due to low liquid holdup.
- **Flexible capacity**: By simply adding or removing plate units, varying evaporation rates may be achieved. Large expansions are possible with additional frames and plate units.

- **Low installation costs**: Due to the compact size and low weight, no cranes or special foundations are required for installation. Typically, an existing building on site is suitable and installation time is minimal.

6.1.5.1.12 Mechanically Aided Evaporators

Mechanically aided evaporators are primarily used for two reasons: (i) to mechanically scrap fouling products from the heat transfer surface; and (ii) to help increase heat transfer by inducing turbulence. Different types are available, including agitated vessels, scraped surface evaporators, and mechanically agitated thin film evaporators (ATFD).

The ATFD system is used in industrial operations for drying of products, drying of concentrated liquid for salt recovery, and drying of chemical and petrochemical products to recover the powder. Through ATFD, water or solvents are evaporated from concentrated liquid to make dry powder or flakes. ATFD is the ideal apparatus for continuous processing of concentrated material to dry solids. It consists of a cylindrical, vertical body with a heating jacket and rotor inside the shell which is equipped with rows and pendulum blades all over the length of the dryer. The hinged blades spread the wet feed product in a thin film over the heated wall. Turbulence increases as the product passes through the clearance before entering a calming zone situated behind the blades as the heat transfers from the jacket to main shell. Under smooth agitation, water or solvent will evaporate and liquid will be converted to slurry, cake, or dry powder or flakes. The vapors produced rise upward, counter-currently to the liquid, and pass through a cyclone separator mounted on the vapor outlet of the ATFD. Furthermore, these vapors will be condensed in the condenser and recovered as condensate. The system is operated under vacuum for temperature-sensitive products and atmospheric conditions for normal drying.

6.1.5.2 Application of MEEs

Evaporators are an integral part of several process industries such as agrochemicals and pesticides, the pharmaceutical sector, pulp and paper, sugar, distillery, dairy, food processing, oil refineries, fertilizer, chloralkali plants, dye and dye intermediate units, and common effluent treatment plants (CETP).

Disposal of effluent from industries is an increasing problem in India and worldwide. The manufacturing process involves usage of more organic and inorganic salts, which is becoming a major part of high chemical oxygen demand (COD) and high total dissolved solids (TDS) in WW. The industry uses MEEs to achieve ZLD for WW handling. RO reject is conveyed to the MEE to separate salt and reusable water. In industrial operations, evaporators are used for some other purposes as follows:

- To reduce water concentration and increase shelf-life of products.
- To reduce packaging, transportation, and storage cost (reduction in bulkiness).
- To improve the stability and handling of products.
- Preconcentration reduces the cost of energy for subsequent processes.
- For the recovery of expensive solvents and to prevent their wastage.

6.1.5.3 Problems Associated with MEEs

Problems associated with an MEE system are that it is an energy-intensive system and therefore any measure to reduce the energy consumption by reducing the steam consumption will help in improving the profitability of the plant. In order to cater to this problem, efforts to propose new operating strategies have been made by many researchers to

Table 6.1 Operation and troubleshooting of MEE system.

Operation problem	Improvement action point
Low vacuum during operation of evaporation system	Check shutoff vacuum
Leakage through the evaporation system	Carry out hydro test
High level in vapor liquid separator (VLS)	Maintain level
Water heating in the ventury tank	Check delta T at cooling tower
High temperature in the calandria	Adjust and manipulate steam supply as per requirement
Nozzle sizing is not perfect at the ventury tank	Check nozzle chocking issue
Chocking of calandria	Evaporation system needs to be flushed and cleaned
Low rate of evaporation	Level increase in calandria, chocking of calandria
Carryover from VLS	Maintain level

minimize the consumption of live steam in order to improve the steam economy of the system. Some of these operating strategies are feed, product, and condensate flashing, feed and steam splitting, and using an optimum feed flow sequence.

Based on operation and troubleshooting of the MEE system, some major points for improvement with respect to vacuum, temperature, and rate of evaporation are mentioned in Table 6.1.

6.1.5.4 Power Consumption

Power consumption is in the range of 22–40 kWh/m^3 of product water produced.

6.1.5.5 Dissolved Solids Removal

The system is generally used to reduce dissolved salts from 50 000 to 750 000 ppm.

6.1.5.6 Capital Cost

In general, total capital cost is in the range of INR 900–3000/m^3 of product water produced.

6.2 ZLD Technologies: Techno-Economic Evaluation

Water pollution by inappropriate management of industrial WW is one of the major environmental problems encountered globally. It is extremely necessary to treat as well as recycle industrial WW to reduce water scarcity and decrease the burden on natural resources. Industrial ZLD has received steadily increasing attention across the world due to stringent regulations and water scarcity. To achieve industrial ZLD, an innovative and sustainable approach is required to select the right technology.

Before considering ZLD, it is important to understand treatment goals, economics, and regulatory requirements. In this section we will discuss techno-economic evaluations of various available ZLD technologies. This techno-economic evaluation approach will help industry to address the issue of minimizing waste as well as help in adoption of cleaner production technologies. In this section techno-economic evaluation is carried out for RO, Scaleban, FO, vacuum distillation, and MEE cum crystallizer.

6.2.1 Power Consumption

Power consumption is one of the important factors in selecting the right ZLD technology. Power consumption range is summarized below for various available ZLD technologies:

- Power consumption in an RO system is in the range of 2.5–5 kWh/m^3 of product water produced.
- Power consumption in a Scaleban system is in the range of 0.5–2 kWh/m^3 of product water produced.
- Power consumption in an FO system is in the range of 7.5–10 kWh/m^3 of product water produced.
- Power consumption in a vacuum distillation system is in the range of 15–22.5 kWh/m^3 of product water produced.
- Power consumption in an MEE cum crystallizer system is in the range of 22–40 kWh/m^3 of product water produced.

Power consumption range associated with various ZLD technologies is shown in Figure 6.1.

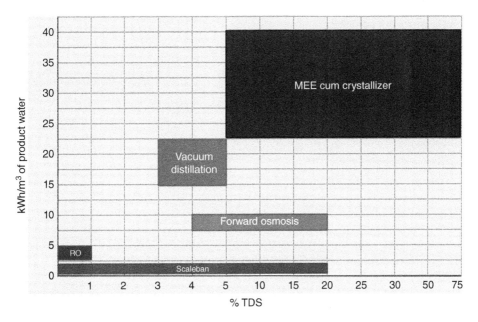

Figure 6.1 Power consumption range of various ZLD technologies.

6.2.2 Capital Cost

Capital cost is one of the important factors in selecting the right ZLD technology. Capital cost range at tertiary stage for WW treatment is summarized below for various available ZLD technologies:

- Capital cost associated with an RO system is in the range of INR 50–100/m³ of product water produced.
- Capital cost associated with a Scaleban system is in the range of INR 150–200/m³ of product water produced.
- Capital cost associated with an FO system is in the range of INR 300–400/m³ of product water produced.
- Capital cost associated with a vacuum distillation system is in the range of INR 700–1500/m³ of product water produced.
- Capital cost associated with an MEE cum crystallizer system is in the range of INR 900–3000/m³ of product water produced.

Capital cost associated with various ZLD technologies is shown in Figure 6.2.

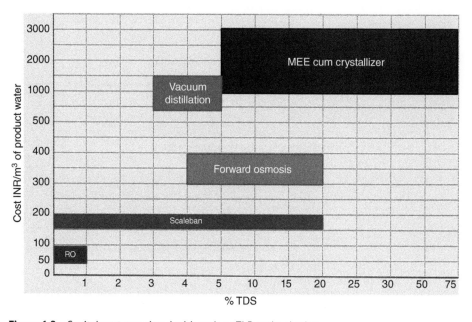

Figure 6.2 Capital cost associated with various ZLD technologies.

6.3 Feasibility Study of ZLD

There are several methods available for the treatment of WW. These may be classified into two categories: to meet standard discharge norms and to meet ZLD norms. We carried out a comparative feasibility study for 2.5 million liters/day (MLD) capacity ETP based on environmental parameters considering power consumption, CO_2 emission, solid waste generation, and chemical consumption, and economic parameters considering capital expenditure and operational expenditure.

6.3.1 Study Methodology

6.3.1.1 Wastewater Quantity and Characteristics
WWTP capacity considered in this case is 2.5 MLD. Untreated WW characteristics and treated WW characteristics are shown in Table 6.2.

Table 6.2 Untreated and treated WW characteristics.

Parameters	Units	Untreated WW characteristics	Treated WW characteristics to meet discharge norms
pH		1.5–11	6.5–7.5
COD	mg/l	2000–5000	<250
BOD	mg/l	500–2500	<100
TSS	mg/l	50–300	<100
TDS	mg/l	2000–13 500	NA
O&G	mg/l	<10	<10
Ammoniacal nitrogen	mg/l	50–100	<50

6.3.1.2 Effluent Treatment Scheme to Meet Standard Discharge Norms
The effluent treatment scheme to meet standard discharge norms is described below and shown in Figure 6.3.

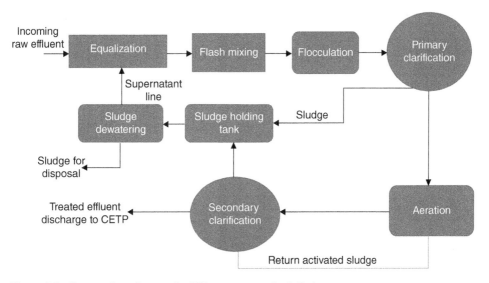

Figure 6.3 Process flow diagram for ETP to meet standard discharge norms.

- **Primary treatment**: This consists of screening, equalization, neutralization, coagulation, flash mixing, flocculation, and primary clarification for removal of suspended particles and some organic impurities.
- **Secondary treatment**: This consists of activated sludge process where effluent will be received after primary treatment. Biological process takes place for removal of organic impurities and further secondary clarification for solid impurities removal.
- **Sludge dewatering**: Sludge dewatering technology is implemented for removal of moisture/excess water from sludge and collected water is sent to an equalization tank for further treatment. The dried cake sludge is sent to authorized landfill sites for disposal.

Treated effluent is discharged into the sea/estuary directly or by CETP, meeting the standard discharge norms.

6.3.1.3 Effluent Treatment Scheme to Meet ZLD Norms
The effluent treatment scheme to meet ZLD norms is described below and shown in Figure 6.4.

- **Primary treatment**: This consists of screening, equalization, neutralization, coagulation, flash mixing, flocculation, and primary clarification for removal of suspended particles and some organic impurities.
- **Secondary treatment**: This consists of activated sludge process where effluent is received after primary treatment. Biological process occurs for removal of organic impurities and further secondary clarification for solid impurities removal.
- **Sludge dewatering**: Sludge dewatering technology is implemented for removal of moisture/excess water from sludge and collected water is sent to an equalization tank for further treatment. The dried cake sludge is sent to authorized landfill sites for disposal.
- **Tertiary treatment**: This consists of removal of residual organic matter, total suspended solids (TSS), and backwash water by tertiary clarifier, filter media, and further treatment by ultrafiltration (UF) and RO for maximum WW recycling; then RO reject is sent to an MEE and further ATFD to achieve ZLD norms. Condensate from both MEE and ATFD is sent to the tertiary clarifier tank. RO permeate is used in the cooling tower and other processes. The dried salt solid waste is sent to authorized landfill sites for disposal.

6.3.2 Feasibility Study Results and Discussion

A techno-economic feasibility study of ZLD plant was carried out based on the following:

- Environmental parameters: power consumption, CO_2 emission, solid waste generation, and chemical consumption.
- Economic parameters: capital expenditure and operational expenditure.
  Results of our study are shown in Table 6.3.

6.3.2.1 Environmental Feasibility
It has been found that power consumption and CO_2 emission to meet ZLD norms are four times higher than to meet standard discharge norms; and that solid waste generation and

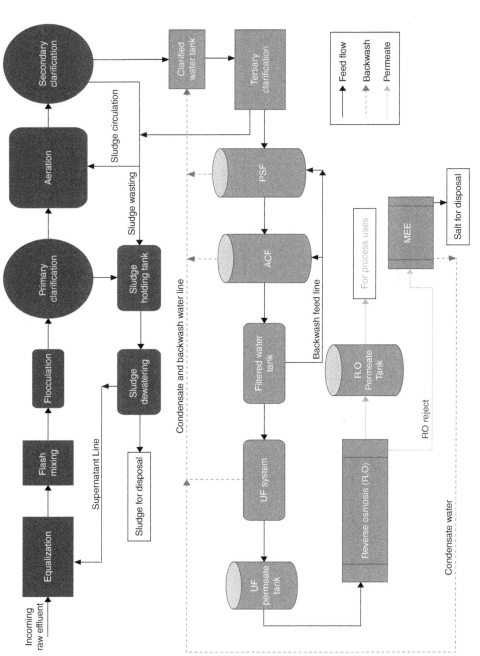

Figure 6.4 Process flow diagram for ETP to meet ZLD norms. PSF, pressure sand filter.

Table 6.3 Feasibility study to meet discharge norms and ZLD norms.

Parameters	Units	ETP to meet discharge norms	ETP to meet ZLD norms	Difference
Power consumption	kW/day	7740	30760	4 times higher in ZLD
CO_2 emission	tons/day	6.3	25.2	4 times higher in ZLD
Solid waste generation	tons/day	30	61	2 times higher in ZLD
Chemical consumption	kg/day	1160	2626	2 times higher in ZLD
Capital cost	INR in Crore	40	100	2.5 times higher in ZLD
Operational cost	INR/kl	500	1500	3 times higher in ZLD

chemical consumption to meet ZLD norms are two times higher than to meet standard discharge norms. See Figure 6.5.

6.3.2.2 Economic Feasibility

It has been found that capital costs to meet ZLD norms are 2.5 times higher than to meet standard discharge norms; and that operational costs to meet ZLD norms are three times higher than to meet standard discharge norms. See Figure 6.6.

6.3.3 Feasibility Study Summary and Conclusions

- The concept of establishment of ZLD may be good but it cannot be generalized and implemented uniformly over the entire country, as it emits more carbon, leading to global warming, and generates huge amounts of hazardous solid waste, occupying more landfill land.
- The Paris Agreement on climate change has identified CO_2 emissions as one of the largest root causes of global warming; hence it is the responsibility of world communities to reduce CO_2 emissions. ZLD carbon emissions are four times more and thus we need to implement ZLD carefully.
- ZLD is good for those areas where suitable treated WW discharge facilities are not available and water scarcity is extreme.
- ZLD is not suitable for coastal areas due to the following reasons:
  - 4 times higher power consumption in ZLD compared to conventional treatment to meet discharge norms.
  - 4 times higher CO_2 emission in ZLD compared to conventional treatment to meet discharge norms.
  - 2 times higher hazardous waste generation in ZLD compared to conventional treatment to meet discharge norms.

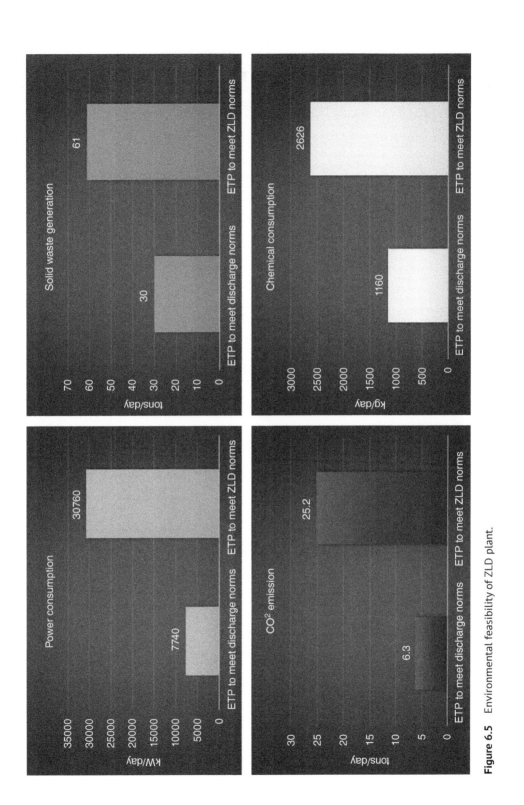

Figure 6.5 Environmental feasibility of ZLD plant.

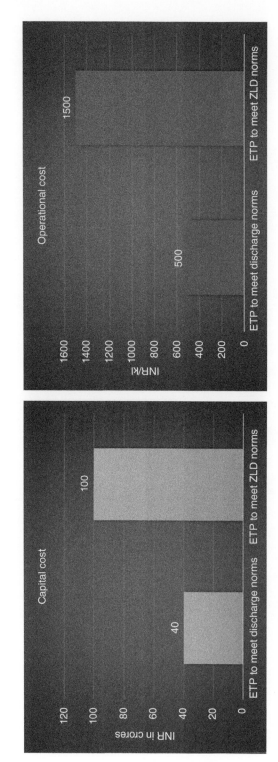

Figure 6.6 Economic feasibility of ZLD plant.

- 2 times higher chemical consumption in ZLD compared to conventional treatment to meet discharge norms.
- 2.5 times higher capital cost in ZLD compared to conventional treatment to meet discharge norms.
- 3 times higher operating cost in ZLD compared to conventional treatment to meet discharge norms.
- Considering the above adverse impacts of ZLD on the environment, it has been suggested to implement ETP to meet standard discharge norms where sea/estuary discharge is possible with the following control measures:
  - Target to reduce freshwater consumption and effluent generation with latest sustainable technologies.
  - Implement effluent stream identification, characterization, and segregation at sources for better effluent management and treatment.
  - Implement latest and advanced technologies for maximum recycling of treated WW to reduce quantity of treated effluent discharge.
- In coastal areas desalinated water is 10–15 times cheaper than ZLD water; hence use of desalinated water is a better choice than ZLD water.
  - Cost of desalinated water: 50–80 INR/m^3.
  - Cost of ZLD water: 750–1200 INR/m^3
- In coastal areas rainwater can be a better choice to be used during the rainy season.
  - Cost of rainwater: <10 INR/m^3.
  - Cost of ZLD water: 750–1200 INR/m^3

6.3.4   My Experience with ZLD

The author of this book is personally involved in the design, installation, commissioning, and operation of several ZLD plants across the world. Below is a case study of a 1000 m^3/day ZLD plant, with actual operating cost and power consumption data. This ZLD plant was designed to treat 1000 m^3/day industrial WW having COD 4000–5000 ppm and TDS 3000–4000 ppm. The treatment units associated with this plant are biological treatment, UF, RO, and MEE cum crystallizer. Industrial WW is first passed through biological treatment where COD is reduced to <250 ppm. After biological treatment WW is passed through UF and RO to reduce the TDS. Permeate of RO is utilized inside the cooling tower, and reject of RO is passed through MEE for further treatment. Almost all water gets recovered with an ZLD plant and reused inside the operation. Sludge from the biological treatment unit and salt from the MEE unit are sent to a secured landfill site for further disposal. Figure 6.7 summarizes the author's experience with ZLD. One-year average data from this plant are summarized below.

6.3.5   Operating Cost

Operating cost includes: chemical cost, power cost, sludge and salt disposal cost, and manpower cost.

- Operating cost of biological treatment unit: 500 INR/m^3 feed effluent (7 USD/m^3).
- Operating cost of UF + RO treatment unit: 136 INR/m^3 feed effluent (2 USD/m^3).

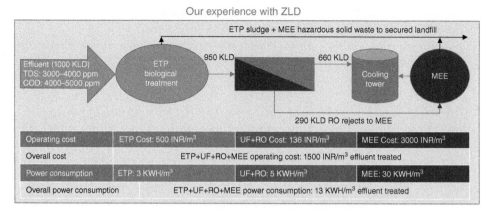

Figure 6.7 Operating cost and power consumption breakup in ZLD plant.

- Operating cost of MEE treatment unit: 3000 INR/m^3 feed effluent (43 USD/m^3).
- Overall operating cost of ZLD plant (ETP+UF+RO+MEE): 1500 INR/m^3 effluent treated (21 USD/m^3).

6.3.6 Power Consumption

Power consumption includes power consumption in power drives of the biological unit, UF+RO unit, and MEE unit. Power consumption in making steam for MEE is also included.

- Power consumption in biological treatment unit: 3 kWh/m^3 feed effluent.
- Power consumption in UF+RO treatment unit: 5 kWh/m^3 feed effluent.
- Power consumption in MEE treatment unit: 30 kWh/m^3 feed effluent.
- Overall power consumption in ZLD (ETP+UF+RO+MEE): 13 kWh/m^3 effluent treated.

My experience with ZLD is not so good that I can recommend ZLD everywhere. ZLD is not an environmently friendly technology where sea discharge options are available and hence should not be mandated for coastal areas. Efforts need to be targeted on reduction of water consumption and WW generation with the latest sustainable technologies. Effluent stream identification, characterization, and segregation at source are the best ways to manage and treat complex industrial effluent. Utilization of rainwater, desalinated water, and treated municipal sewage water are better choices than ZLD water for industries and communities. However, ZLD may be recommended for those areas where suitable treated WW discharge facilities are not available and water scarcity is extreme.

Further Reading

American Public Health Association (1985). *Standard Methods for the Examination of Wastewater*, 16e. Washington, DC: APHA, AWWA, WPCE.

Birdi, G.S. and Birdi, J.S. (1988). *Water Supply and Sanitary Engineering*. New Delhi: Dhanpat Rai Publishing Company.

Chaubey, M. (2002). Computer design of wastewater treatment plant using fixed film bioreactor technology. *Journal of Industrial Pollution Control* 18 (1): 107–117.

Chaubey, M. and Kaushika, N.D. (2003). Performance analysis of fixed film bioreactor plants for wastewater treatment. *Journal of Industrial Pollution Control* 19 (2): 203–213.

Kaushika, N.D. and Chaubey, M. (2002). Energy efficiency in wastewater treatment plants. *Environment Pollution Control Journal*: 37–40.

Metcalf & Eddy Inc (2014). *Wastewater Engineering: Treatment and Resource Recovery*, 5e. New Delhi: Tata McGraw-Hill.

Rudolfs, W. (1953). *Industrial Waste Treatment*. New York: Reinhold Publishing Corporation.

7

Wastewater Treatment Plant Operational Excellence and Troubleshooting

Wastewater treatment plant (WWTP) operation and troubleshooting is a challenging and interesting field. The success of any WWTP depends on how we operate the plant. The process needs to be operated and stabilized based on microbial activity in addition to physical and chemical parameters that take time in acclimatization with regard to incoming wastewater (WW). However, chemical reaction is based on various operating physical and chemical parameters.

7.1   Wastewater Treatment Issues

The major WW treatment issues faced by industries are summarized below:

- Changing streams:
  - Very complex.
  - With high total dissolved solids (TDS) stream.
  - Highly variable flow.
  - Highly variable quality.

Wastewater Treatment Technologies: Design Considerations, First Edition. Mritunjay Chaubey.
© 2021 John Wiley & Sons Ltd. Published 2021 by John Wiley & Sons Ltd.

- Performance problems:
  - That result in production constraints.
  - Create compliance issues.
- Merging all streams into one common effluent treatment plant (ETP).
- Poor performance due to lack of stream identification and segregation.
- Lack of advanced and latest sustainable technology use.
- Higher environmental footprint of WWTPs.
- Higher operating cost.
- Any time any kind of effluent gets added into the ETP without impact assessment.

7.2 Wastewater Stream Identification, Characterization, and Segregation

A structured approach toward WW treatment helps us to manage complex industrial efflu-ent. The best way to manage complex and variable industrial WW is by WW stream identifi-cation, characterization, and segregation. The incoming effluent stream should be identified and segregated into green, yellow, and red streams. The green stream may consist of all the WW stream having TDS < 5000 ppm and COD < 10 000 ppm. The yellow stream may consist of all the WW stream having 5000 < TDS < 100 000 ppm and COD < 20 000 ppm. The red stream may consist of all the WW stream having TDS > 100 000 ppm and COD > 20 000 ppm. After stream identification and segregation, the green stream may be treated with biological treatment technologies such as activated sludge process (ASP) or moving bed biofilm reac-tor (MBBR). The yellow stream may be treated with forward osmosis (FO), Scaleban, or OH radical technology. The red stream may be treated with multiple-effect evaporation (MEE) technology or any other appropriate evaporation technology. Normally the quantity of the green stream is higher and the quantity of the red stream is minimal. Stream identification, characterization, and segregation help us to optimize the cost of WW treatment and enable us to manage large and complex effluent treatment in an efficient way (see Figure 1.5).

7.3 Operation and Troubleshooting for Preliminary Treatment System

7.3.1 Screening

Raw effluent passes through one or two stages of coarse and fine screening. This can be automatic and/or manual based on operational requirement. The main purpose of screening is to remove the large floating material which may damage downstream equipment. The angle of inclination of the screens is generally 45° to the horizontal. Screens must be cleaned on a regular basis.

7.3.2 Wastewater Equalization

The incoming WW is to be equalized before transfer for treatment further downstream. The main purpose of equalization is to maintain the pH as per the downstream feed requirement, and flow and organic loading is balanced in such a way that downstream shock loading (feed fluctuations) does not happen. For equalization of WW, two receiving tanks should be considered, one for receipt of incoming effluent and the other for processing in downstream treatment. The offspec effluent is received into an offspec tank. For proper mixing of incoming effluent, a blower with diffuser arrangement is provided. Normally the offspec tank is kept empty, so that when offspec effluent is reported it can be easily accommodated. The operation of effluent transfer pumps, blowers, and monitoring instruments connected with the equalization tank must be done in accordance with standard operating procedures.

7.3.3 Wastewater pH Neutralization

Downstream treatment generally requires a neutral pH in the range of 6.5–8.0 for best performance of the treatment system. The incoming WW may have a lower or higher pH, so the pH must be monitored with either an online pH meter or pH paper.

7.4 Operation and Troubleshooting for Primary Treatment System

The basic purpose of the primary treatment system is to remove total suspended solids (TSS), oil and grease (O&G), and some specific pollutants. In some effluent streams the removal of sulfides, heavy metals, total nitrogen (free ammonia removal by stripping and precipitation), and phosphorus (particulate phosphorus removal by membrane separation) is required. The chemical process is either charge neutralization and/or the sweep floc mechanism. During this destabilization of colloids, agglomeration of flocs occurs. Various types of chemicals are used such as alum, lime, ferric salts for coagulation, and polymer dosing for further flocculation. For clarification of coagulated and flocculated types of impurities, different types of clarification mechanisms are operated such as conventional primary clarifier, lamella and tube settler unit, dissolved air flotation (DAF), pipe flocculator unit, and high rate solid contact clarifier (HRSCC) unit. O&G removal treatment units like American plate interceptor (API), tilted plate interceptor (TPI), and DAF are used based on the type and concentration of O&G.

The incoming equalized WW is pumped into primary treatment units. Some of the operation and troubleshooting points are mentioned below:

- Chemical doses should be optimized based on a jar test, for better removal of TSS and O&G.
- Chemical sludge wasting should be carried out and monitored on every shift.
- Incoming pH should be maintained near 7.0–8.0, for better coagulation and flocculation.
- The air flow rate and recycling ratio of the WW should be monitored during DAF unit operation.
- Proper operation of all mechanical equipment that is submerged in WW should be ensured, such as flash mixer, flocculation mixer, and clarification mechanism.

7.5 Operation and Troubleshooting for Secondary Biological Treatment System

The basic purpose of the secondary biological treatment system is to remove the organic load from the effluent. In some specific effluent streams, removal of nitrogen and phosphorus (organic compound forms) is also required. The chemical process is that of microbiological growth, where organic pollutants are converted into CH_4, CO_2, H_2O, and nitrogen compounds through simultaneous nitrification and denitrification (SND) converted into nitrogen (N_2). Phosphorus is consumed during biological growth as bio-absorption.

7.5.1 Aerobic Biological Treatment Units

Based on requirement and the nature of the raw effluent, many types of aeration system are installed and operated:

- Conventional aeration: Biological growth occurs in suspended forms only.
- Sequencing batch reactor (SBR): A form of suspended biological growth. Designed for batch operation as fill–aerate–settle mode; with the help of a specific decanter mechanism, treated WW is decanted from the aeration tank.
- MBBR: Biological growth occurs in both attached and suspended forms. Plastic media are used for growth of bacteria.
- MBR: This is a highly advanced technology, being operated with very high mixed liquor suspended solids (MLSS) of approx. 6000–10000 ppm. It is mostly used where treated WW is being recycled. Permeate from the MBR system is fed directly into the reverse osmosis (RO) system.

For secondary biological treatment units, some operation and troubleshooting points are mentioned below:

- Feed pH to be maintained near 6.5–7.5 in view of better performance of biological system.
- MLSS to be maintained more than 3000 ppm as per technology uses.
- Dissolved oxygen (DO) in aeration tank to be maintained in the range of 2–3 ppm.
- Food to microorganism (F/M) ratio to be maintained in the range of 0.15–0.3 based on incoming effluent nature.
- Biological sludge wasting to be monitored and carried out in every shift.
- Temperature inside biological aeration tank must be maintained in range of 20–30 °C. Never allow temperature to go below 5 °C or above 35 °C.
- Ensure proper operation of all mechanical equipment that is submerged in WW such as diffusers and secondary clarification mechanism.

7.5.1.1 Aeration Volume and Oxygen Requirement

Oxygen is required by microorganisms for metabolic activity, and in the aerobic process oxygen is at the heart of the system. Various types of air supply system have been developed over time in view of the oxygen transfer rate (OTR). The main types are:

- Surface aeration.
- Coarse bubble aeration.
- Surface jet aeration.
- Fine bubble aeration: Based on type of membrane (ethylene propylene diene monomer [EPDM], polyurethane, or silicon), elasticity, and O_2 transfer – number of pores varies in same areas (losing elasticity leads to clogging and damage during operation).

Basic aeration tank volume, oxygen requirement, and blower capacity calculations based on COD are shown in Table 7.1.

Table 7.1 Aeration tank volume, air requirement, and blower capacity calculations.

Description	Units	Value
Aeration tank volume		
COD at the inlet in aeration tank (assumed)	mg/l	5000
Flow (assumed)	cum/h	100
	cum/day	2400
Total COD load in aeration tank	kg/day	12000
F/M ratio at COD (COD basis)		0.3
MLSS (3000–4500) (normal aeration system)		3500
Aeration tank volume based on F/M ratio	**cum**	**11429**
Air requirement		
Oxygen requirement (0.8–1.4 kg O_2/kg of COD)	kg/kg	1.2
COD in aeration system	kg/day	12000
Oxygen required under field condition	kg/day	14400
Residual DO in aeration tank (1.5–2.5)	mg/l	2
Oxygen in aeration tank (DO)	kg/day	17
Total oxygen needed	kg/day	14417
Air (specific weight or density) @ 37 °C (1.4–1.8)	kg/cum	1.139
Content of oxygen in air	%	0.21
Air requirement	cum/day	60275
Transfer efficiency maximum value (from 15 to 30% based on type of aeration system). Here considered for fine bubble system	%	0.24
Air requirement	**cum/day**	**251145**
Blower capacity		
Air needed	cum/h	10464
No. of working blowers considered		4 (4 working, one standby)
Blower capacity	cum/h	2616
Total blower capacity	**cum/h**	**10464**

7.5.1.2 Sludge Recycling and Wasting for MLSS

Any biological system works on the basis of microbial biomass content balanced with the sludge recycling and wasting system. MLSS in a biological aeration system depends on the type of bioreactor system. Once the required MLSS is achieved, sludge wasting will be required from the bioreactor. Table 7.2 shows sludge production calculations.

Sludge wasting should be carried out on a continuous basis from the bioreactor. If not possible due to any reason, then it should be done in two to three hours every shift on a daily basis based on the downstream sludge handling system.

7.5.1.3 Food to Microorganism Ratio

For better performance of any biological system, a balance between organic loading and microbial biomass content should be maintained in the bioreactor system. The food to

Table 7.2 Sludge production calculations from an aeration system.

Aerobic sludge – COD basis	Value	Units
Inlet COD in ASP	3500	ppm
Outlet COD from clarifier	800	ppm
Flow rate	1000	m³/day
COD destruction in ASP	2700	kg/day
COD sludge yield factor	0.3	kg DS/kg COD destroyed
COD sludge generated	810	kg/day
Total sludge generated	810	kg/day
Sludge consistency	1.2	%
Sludge generated from aeration tank	67.5	m³/day
Total wet sludge	**2.81**	**m³/h**

Table 7.3 F/M ratio calculations.

Description	Units	Value
COD at the inlet	mg/l	3500
Flow	cum/h	50
	cum/day	1200
Food incoming	kg/day	4200
MLSS	mg/l	3500
Volume of aeration tank	cum/day	4000
F/M ratio calculated (Food / MLSS × Volume)		**0.3**

microorganism (F/M) ratio is one of the significant design and operational parameters of activated sludge systems. A balance between substrate consumption and biomass generation helps in achieving system equilibrium. The F/M ratio is responsible for the decomposition of organic matter. The type of activated sludge system can be defined by its F/M ratio.

- Extended aeration, $0.05 < F/M < 0.15$.
- Conventional activated sludge system, $0.2 < F/M < 0.5$.
- Completely mixed, $0.2 < F/M < 1.0$.
- High rate, $0.4 < F/M < 1.5$.

Table 7.3 shows F/M ratio calculations.

7.5.1.4 Relationship of TOC, COD, and BOD

Chemical oxygen demand (COD) and biological oxygen demand (BOD) analysis is required sometimes. In view of daily operations of a WWTP, total organic carbon (TOC) to COD ratio should be established and operated based on online TOC analyser. Ratio varies from two to five based on effluent nature.

7.5.1.5 Temperature Rise in Biological Reactor

In several WWTPs it has been found that temperature increases in the biological reactor. Operators should know that when the temperature of the bioreactor is too high (above 35 °C) it can have an adverse impact on the microorganisms in all types of biological treatment processes including ASP, MBBR, and MBR.

7.5.1.6 Basic Guidance

The biological system is a very sensitive and self-sustaining process. There are many reasons or conditions why biological systems my be disturbed and affected, including sudden increases in organic loading, changes in characteristics of WW, fluctuations in flow, and changes in operating conditions (pH, DO, temperature, variation in nutrient ratio N:P:K).

The biological system is very much dependent on microbial activity, growth, and acclimatization of microbes in the conditions. The properties of microbes are also limited to the biological treatment, for example sudden changes in temperature or low availability of oxygen in the biological tank may disturb microbial activity and adversely affect performance of the plant.

7.5.1.7 Details of Bacterial Temperature Ranges

Generally, bacteria fall into one of three temperature classifications or ranges, as shown in Table 7.4. This adaptability of microorganisms is stable up to a certain temperature range (35–40 °C); after this, bacterial performance and adaptability are affected, and it may also lead to a diminished bacterial colony.

Table 7.4 Temperature classification of biological processes.

Type of bacteria	Temperature range (°C)	Optimum temperature range (°C)
Psychrophilic	10–27	12–18
Mesophilic	22–45	27–40
Thermophilic	35–75	55–60

7.5.1.8 Root Cause of Temperature Rise in Biological Reactor

Temperature affects the performance of the biological bioreactor as a result of its impact on the rates of biological reactions. Two additional factors must be considered: the maximum acceptable operating temperature and the factors that affect heat loss and gain in the process.

The maximum acceptable operating temperature for biological systems/bioreactor is limited to about 35–40 °C (95–104 °F), which corresponds to the maximum temperature for the growth of mesophilic organisms. Even short-term temperature variations above this range must be avoided, since thermal inactivation of mesophilic bacteria occurs quickly. Successful operation can also be maintained if temperatures are reliably kept above about 45–50 °C (113–122 °F), since a thermophilic population will develop, provided that thermophilic bacteria exist with the capability to degrade WW constituents. Unacceptable performance will result with temperatures between about 40 and 45 °C due to the limited number of microorganisms that can grow within this range. These considerations are particularly important for the treatment of high-temperature industrial WWs.

One of the factors that affects heat gains in biological processes is the production of heat as a result of biological oxidation. The growth of bacteria requires that a portion of the electron donor be oxidized to provide the energy needed for biomass synthesis. Energy is also needed for cell maintenance. This oxidation and subsequent use of the energy result in the conversion of that energy into heat. Although this may seem surprising at first, it is directly analogous to the release of energy that occurs when material is burned; the only difference is the oxidation mechanism. The amount of heat released in the bio-oxidation of carbonaceous and nitrogenous material is directly related to the oxygen utilized by the process. For each gram of oxygen used, 3.5 kcal of energy are released. Since 1 kcal is sufficient energy to raise the temperature of 1 l of water 1 °C, the impact of this heat release depends on the WW strength. For example, a typical domestic WW requires only 1 g of oxygen for each 10 l treated; therefore the temperature rise would be only 0.35 °C, a negligible amount. On the other hand, it is not unusual for an industrial WW to require 1 g of oxygen for each liter treated, in which case the temperature rise would be 3.5 °C. This could be quite significant, particularly if the WW itself is warm.

WW temperature may also increase due to the sulfide concentration in WW. As the bio-oxidation starts in the bioreactor, temperature may increase from 6 to 8°C when sulfide concentration is in the range of 300–800 ppm. This problem can be resolved with controlled dosing of ferrous sulfate ($FeSO_4$) at the pretreatment stage. By adding $FeSO_4$ the sulfide concentration decreases and the chances of increasing temperature in the bioreactor reduces during bio-oxidation.

Other heat gains and losses occur in biological systems. Heat inputs to the system include heat of the influent WW, solar inputs, and mechanical inputs from the oxygen transfer and mixing equipment. Heat outputs include conduction and convection, evaporation, and atmospheric radiation. The main causes of a temperature rise in a biological reactor are summarized below:

- Oxidation of organic matter.
- Influent WW temperature.
- Air blower temperature.
- Atmospheric temperature.
- Sulfide content.

7.5.2 Typical Operating Parameters in an Activated Sludge Process

In an activated sludge process, based on effluent characteristics we may select any of following units or combination of units:

- Activated sludge unit (ASU).
- Denitrification/nitrification biotreater (DNB).
- Intermittent nitrification/denitrification biotreater (INDB).

Typical operating parameters in the ASU, DNB, and INDB are shown in Table 7.5.

Table 7.5 Typical operating parameters for ASU, DNB, and INDB.

Parameter	ASU	DNB	INDB
General			
Temperature (°C)	20–35	20–35	20–35
F/M (day^{-1})	0.15	0.15	0.15
MLSS – reactor (g/l)	4	4	4
MLVSS – reactor (g/l)	3	3	3
MLSS – bleed (g/l)	10–15	10–15	10–15
Sludge recycle ratio	1	1	1
Mixed liquor recycle ratio	–	3	–
Nitrification time (min/h)	–	–	40–50
Denitrification time (min/h)	–	–	10–20
Aerobic sludge age (days)	20–30	20–30	20–30
P required (kg/250 kg BOD)	1	1	1
COD/BOD in	2–2.5	2–2.5	2–2.5
COD/N reactor	>5	>5	>5
BOD-biomass (kg/kg)	0.6	0.6	0.6
Contact zone (selector)			
Contact time (minutes)	5–15	5–15	10–15
Anoxic zone			
pH	–	7–8	–
DO (ppm)	–	<0.5	–
Mixing intensity (W/m^3)	–	20	–
Hydraulic retention time (h)	–	2	–
Aeration tank			
pH	7–8	7–8	7–8
DO (ppm)	>0.5; typically 2	>0.5; typically 2	>0.5; typically 2
Mixing intensity (W/m^3)	>20	>20	>20
Hydraulic retention time (h)	10–15	10–15	10–15
Degasifier			
Contact time (minutes)	10–20	10–20	10–20
Clarifier			
Surface load (m/h)	0.5–0.7	0.5–0.7	0.5–0.7
Hydraulic retention time (h)	2–3	2–3	2–3
Sludge stabilization tank[a]			
Storage capacity (days)	7–14	7–14	7–14
Chemical dosing[b]	H_3PO_4	H_3PO_4	H_3PO_4
	Methanol/ethanol	Methanol/ethanol	Methanol/ethanol
	PE	PE	PE
	NaOH	NaOH	NaOH
	Trace metals	trace metals	Trace metals

[a] A sludge stabilization tank is foreseen to have seed sludge available after a calamity has occurred.
[b] H_3PO_4 is dosed as P-source. NaOH is dosed as pH control. Methanol or ethanol is dosed only if additional COD for denitrification is needed. PE is dosed only when sludge settling properties are insufficient. Trace metals are rarely dosed; this is only done when a shortage has been observed.

7.6 Operation and Troubleshooting for Tertiary Treatment System

The basic purpose of a tertiary treatment system is to polish the treated effluent for further recycle, reuse, or discharge. In some specific effluent streams, removal of heavy metals is required before feeding into ultrafiltration (UF) and RO systems. The chemical dosing systems required are coagulant dosing, polyelectrolyte dosing, sodium hypochlorite dosing, UF chemically enhanced backwash (CEB)-1 dosing, UF CEB-2 dosing, sodium metabisulphite (SMBS) dosing, antiscalant dosing, sodium hydroxide dosing, and acid dosing, all based on the incoming effluent characteristics.

Based on requirement and purpose of the treated WW, common tertiary treatment systems generally have the following operating units:

- **Media filtration**: The treated water from secondary treatment will not be pure and may contain some very fine suspended particles and bacteria. To remove or reduce the remaining impurities still further, the water is filtered through a bed of fine granular sand. This can be a combination of sand media, activated carbon media, and glass media.
- In some specific cases, automatic self-cleaning filters are also used to screen particulate matter.
- **UF membrane**: To protect the downstream RO system from incoming impurities in WW, a UF system can be installed to remove colloidal silica with other impurities. UF membranes can be constructed from polyvinylidene fluoride (PVDF), polysulfone, or ceramic materials.
- **RO**: For removal of inorganic impurities and maximum recycle/reuse of treated WW, an RO system is required. RO membrane is generally constructed from polyamide, but in some cases with a synthetic coating, like with a disc RO membrane to handle recalcitrant non-reactive COD.

Some operational and troubleshooting points are shown in Table 7.6.

7.6.1 Media Filtration Treatment Unit

For good operational practices and proper functioning of various types of media filtration, flow velocity guidelines are given in Table 7.7. Some other points to be considered during operation of various types of media filtration are that if differential pressure across the filter is more than $0.8 \, \text{kg/cm}^2$ then the filter should be backwashed. At some intervals, sand media should be checked for healthiness with regard to biological growth, size of media, and strainer condition.

7.6.2 Ultrafiltration Treatment Unit

UF is where a porous membrane is used to separate or reject colloidal and particulate matter. Similar to other membrane processes, UF is pressure driven. High permeability of the UF membrane and negligible osmotic effects allow the UF process to operate at relatively low pressures.

Table 7.6 Troubleshooting for aerobic systems.

Observation	Analysis and cure
Poor reactor performance – high COD, BOD, N	Check all operational parameters, check for possible toxicity and biodegradability of WW, check pollutant loading. Take measures according to findings. If sludge is intoxicated, reseeding may be required.
Sludge bulking	Check sludge volume index (SVI), perform microscopic observation, check pollutant loading and sludge concentration, DO concentration, and nutrient levels. Long term: restore normal operational parameters and add nutrients or trace elements if needed. Short term: dosing of chlorine, H_2O_2, $AlCl_3$, $FeCl_3$, or PE may be considered.
Dispersed floc	Check pollutant loading and sludge concentration. Overloading is the most probable cause. Long term: restore normal loading (F/M). Short term: dose of PE.
Deflocculated sludge	Check pollutant loading and sludge concentration. Underloading is the most probable cause. Long term: restore normal loading (F/M). Short term: dose of PE.
Floating sludge	Probable cause is denitrification in clarifier. Decrease sludge retention time in clarifier. Consider introducing denitrification in biotreater.

Table 7.7 Flow velocity for sand filter and activated carbon filter.

Type of service	Flow velocity (m/h)		
	Service flow	Backwash flow	Air scouring flow
Sand media based	10–15	24–36	35–40
Activated carbon media based	15–20	10–15	Not required

7.6.2.1 Basket Strainers

After media filtration, basket strainers trap any sand particles that may leak out or go into the UF unit and avoid permanent chocking of UF fibers. They provide a guard filter for the UF system. Since the frequency of leakage of sand is very low, generally manual backwash is carried out once a month or two. However, valves are provided for backwashing of basket strainers. Backwash waste is routed to the waste disposal tank. Modes of operation of UF are processing, backwash, and chemical cleaning:

- **Processing**: Processing is the mode in which filtration takes place. In dead-end filtration, all water fed to the UF module passes through the membrane and is filtered. During the processing step, dirt material is accumulated on the membrane surface.
- **Backwash**: During processing, solids accumulate on the membrane surface and short cleaning or backwash with filtrate water is necessary to restore membrane performance.

The system is programmed to backwash individual blocks at fixed periodical intervals. During backwash the water flow will be reversed. Clean permeate water will be forced from outside the capillary and will transport the foulant out of the capillaries. Duration of the backwash is enough to remove all dirt from the membrane surface and drive it out from the capillaries and membrane system.

- **Chemical cleaning**: If the backwash is insufficient to remove all the foulants, backwash efficiency can be enhanced by adding chemicals to the backwash (CEB). During CEB it is important that first backwash without chemicals will remove most of the foulants prior to backwash with added chemicals. In this way, chemicals are used only for hard-to-remove problems. It is also important that all fibers are filled with chemicals in the correct amount and with correct soaking time. After soaking it is important to flush all chemicals out with a backwash.

Some of the operational and troubleshooting points are as follows:

- Processing is the mode in which UF takes place. During this operation, the service inlet XV-valve and service outlet XV-valve are automatically opened and will be closed at a prerequisite time for backwash operation.
- Backwash is necessary to restore membrane performance. UF permeate water should be used for backwashing of UF membranes. Backwash operation (for 10 seconds) is carried out after 30 minutes of processing cycle operation based on trans-membrane pressure (TMP). During this operation, backwash pumps will start, service inlet and outlet auto XV-valves will close, and B/W inlet and outlet auto XV-valves will open.
- Chemically enhancing UF permeate water used for backwash operation is done to remove dirt attached to membrane surface with the help of chemicals. Two chemicals enhance backwash (CEB-1 and CEB-2) and this will take place alternately, each system comprising one tank and two pumps, provided for chemically enhancing the backwash water. CEB will be carried out after 32 cycles of filtration/once a day when there is a plant shutdown or when differential pressure reaches $1.2\,kg/cm^2$.
- A soaking step is also required when CEB is employed. Chemically enhanced water is introduced to membranes and soaking for 10 minutes is done by closing all the valves and backwash pumps. This step is provided to allow reaction of chemical with accumulated dirt and further help loosening and cleaning of membranes. After soaking again, membranes are backwashed with clean UF permeate water.

7.6.3 Reverse Osmosis Treatment Unit

RO is a natural process involving fluid flow across a membrane which is said to be "semipermeable." A semipermeable membrane is selective in that certain components of a solution, usually the solvent, can pass through, while others, usually the dissolved solids, cannot. Direction of solvent flow is determined by its chemical potential, which is a function of pressure, temperature, and concentration of dissolved solids.

7.6.3.1 Membrane Cleaning

RO membranes are widely used for salty water treatment. The use of RO membranes in water desalination and WW reclamation, reuse, and recycling has increased over the past few years. However, a major impediment in the use of RO membranes for water desalination and WW reclamation is membrane fouling. During normal operation over a period of time, RO membrane elements are subject to fouling by suspended, organic/inorganic materials that may be present in the feed water. Common examples of foulants are:

- Calcium carbonate scale.
- Sulfate scale of calcium, barium, or strontium.
- Metal oxides (iron, manganese, copper, nickel, aluminum, etc.).
- Polymerized silica scale.
- Inorganic colloidal deposits.
- Mixed inorganic/organic colloidal deposits.
- Organic material (natural organic matter [NOM]).
- Man-made organic material (e.g. antiscalant/dispersants, cationic polyelectrolytes).
- Biological matter (bacterial bioslime, algae, mold, or fungi).

The entry of these foulants in RO systems causes fouling of membranes. The fouling leads to increase in the differential pressure from feed to concentrate and finally leads to membrane flux declination and mechanical damage of the membrane. Foulant removal through chemical cleaning is therefore a major objective of the membrane chemical cleaning process.

7.6.3.2 Chemical Cleaning Requirement

Membranes of RO systems must be cleaned with suitable chemicals in any of the following circumstances:

- When the normalized permeate flow drops by 10%.
- When the normalized salt passage increases by 5%.
- When the normalized differential pressure increases by 15%.

RO chemical cleaning frequency will vary by site. A rough rule of thumb as to an acceptable frequency is once every three to five months. It is important to clean the membranes when they are still only lightly fouled, rather than when heavily fouled.

7.6.3.3 Chemical Cleaning Sequence

Normally three kinds of RO chemical cleaning are recommended: acid cleaning, alkaline cleaning, and sanitizing. The recommended sequence of RO chemical cleaning is given below:

- Acid cleaning and flushing.
- Alkaline cleaning and flushing.
- Sanitizing and flushing.

7.6.3.4 Cleaning Chemicals

Acid cleaners and alkaline cleaners are the two standard cleaning chemicals. Acid cleaners are used to remove inorganic precipitates including iron, while alkaline cleaners are used to remove organic fouling including biological matter. It is recommended that RO permeate should be used for cleaning solutions. Table 7.8 lists suitable RO cleaning chemicals along with recommended concentrations.

Table 7.8 Recommended RO cleaning chemicals.

Chemical	Concentration by weight (%)	Foulant
Sodium hydroxide (NaOH)	0.1	Biological and organic
Tetrasodium salt of ethylene diamine tetra acetic acid (Na_4EDTA)	1.0	Biological and organic
Sodium salt of dodecylsulfate (Na-DDS)	0.025	Biological, organic, and silica
Hydrochloric acid (HCl)	0.2	Inorganic
Sodium hydrosulfite ($Na_2S_2O_4$)	1.0	Inorganic and iron
Phosphoric acid (H_3PO_4)	0.5	Inorganic and iron
Sulfamic acid (NH_2SO_3H)	1.0	Metal oxides
Hydrogen peroxide	0.25	Bacterial (*E. coli*)

7.6.3.5 Flow and Pressure of Chemical Solutions

Flow rate and pressure of circulated chemical solutions during chemical cleaning of RO membranes play very important roles. Table 7.9 lists requirements.

Table 7.9 Recommended flow and pressure of RO chemical cleaning solution.

Membrane diameter (inches)	Flow rate per pressure vessel (m^3/h)	Pressure (bar)
2.5	0.7–1.2	1.5–3.5
4	1.8–2.3	1.5–3.5
8	6.0–9.0	1.5–3.5

7.6.3.6 Temperature and pH Range of Chemical Solutions

Temperature and pH of circulated chemical solutions during chemical cleaning of RO membranes play very important roles. Table 7.10 lists requirements.

Table 7.10 Recommended temperature and pH range of RO chemical cleaning solution.

Temperature (°F)	Operating pH range	Cleaning pH range
20–120	3–11	2–12

7.6.3.7 Chemical Cleaning Equipment

RO chemical cleaning systems consist of a mixing tank, cleaning pump, and micron cartridges. The successful cleaning of an RO on-site requires well-designed RO cleaning equipment. It is recommended to clean a multistage RO one stage at a time to optimize cross-flow cleaning velocity. The source water for chemical solution make-up and rinsing should be clean RO permeate or DI water and be free of hardness, transition metals (e.g. iron), and chlorine. Components must be corrosion proof. The general arrangement of RO chemical cleaning equipment is shown in Figure 4.6. Table 7.11 lists size and material of construction of RO chemical cleaning equipment.

Table 7.11 Size and material of construction of RO chemical cleaning equipment.

Equipment	Size	Material of construction
Mixing tank	20% more to the volume of pressure vessel	HDPE/PP/FRP
Cleaning pump	Flow as recommended in Table 7.2 at pressure 1.5–3.5 bar	SS-316/PP
Cartridge filters	10 μm	–

HDPE, High-density polyethylene; PP, polypropylene; FRP, fibre reinforced polymer; SS, stainless steel.

7.6.3.8 Chemical Cleaning and Flushing Procedure

In an RO plant the RO membrane can be cleaned in place (CIP) in the pressure tubes by recirculating the cleaning solution across the high-pressure side of the membrane at low pressure and relatively high flow. The cleaning process is described below:

- Flush the pressure tubes at low pressure (3.5 bar or 50 psi) by pumping clean water (RO permeate or DI quality and free of hardness, transition metals, and chlorine) from the cleaning tank.
- Mix a fresh batch of the selected cleaning solution in the cleaning tank. The dilution water should be clean water of RO permeate or DI quality and be free of hardness, transition metals, and chlorine. The temperature and pH should be adjusted to their target levels.
- Circulate the cleaning solution through the pressure tubes for approximately one hour or the desired period of time. At the start, send the displaced water to drain so you do not dilute the cleaning chemical and then divert up to 20% of the most highly fouled cleaning solution to drain before returning the cleaning solution back to the RO cleaning tank. For the first five minutes, slowly throttle the flow rate to one-third of the maximum design flow rate. This is to minimize the potential plugging of the feed path with a large amount of dislodged foulant. For the second five minutes, increase the flow rate to two-thirds of the maximum design flow rate, and then increase the flow rate to the maximum design flow rate. If required, readjust the pH back to the target when it changes more than 0.5 pH units.
- An optional soak and recirculation sequence can be used, if required. The soak time can be from one to eight hours depending on the manufacturer's recommendations. Caution should be used to maintain the proper temperature and pH. Also note that this does increase the chemical exposure time of the membrane.

- Upon completion of the chemical cleaning steps, a low-pressure cleaning rinse with clean water (RO permeate or DI quality and free of hardness, transition metals, and chlorine) is required to remove all traces of chemical from the cleaning skid and RO skid. Drain and flush the cleaning tank; then completely refill the cleaning tank with clean water for the cleaning rinse. Rinse the pressure tubes by pumping all of the rinse water from the cleaning tank through the pressure tubes to drain. A second cleaning can be started at this point, if required.
- Once the RO system is fully rinsed of cleaning chemical with clean water from the cleaning tank, a final low-pressure clean-up flush can be performed using pretreated feed water. The permeate line should remain open to drain. Feed pressure should be less than 60 psi (4 bar). This final flush continues until the flush water flows clean and is free of any foam or residue of cleaning agents. This usually takes 15–60 minutes. The operator can sample the flush water going to the drain for detergent removal and lack of foaming by using a clear flask and shaking it. A conductivity meter can be used to test for removal of cleaning chemicals, such that the flush water to drain is within 10–20% of the feed water conductivity. A pH meter can also be used to compare the flush water to drain to the feed pH.
- Once all the stages of a train are cleaned, and the chemicals flushed out, the RO can be restarted and placed into a service rinse. The RO permeate should be diverted to drain until it meets the quality requirements of the process (e.g. conductivity, pH).

7.7 Wastewater Sampling

7.7.1 Sampling

The objective of sampling is to collect a portion of effluent small enough in volume to be transported conveniently and handled in the laboratory while still accurately representing the material being sampled. This implies that the relative proportions or concentration of all pertinent components will be the same in the samples as in the material being sampled, and that the sample will be handled in such a way that no significant changes in composition occur before the tests are made.

The samples should be handled in such a way that it does not deteriorate or become contaminated before it reaches the laboratory. Before filling, rinse the sample bottle out two or three times with the water being collected, unless the bottle contains a preservative or dechlorinating agent. Depending on analyses to be performed, fill the container fully (most organics analyses) or leave space for aeration, mixing, etc. (microbiological analyses).

Representative samples of some sources can be obtained only by making composites of samples collected over a period or at many different sampling points. Make a record of every sample collected and identify every bottle, preferably by attaching an appropriately inscribed tag or label.

7.7.1.1 Grab Samples

A sample collected at a time and place can represent only the composition of the source at that time and place. However, when a source is known to be fairly constant in composition over a considerable period of time or over substantial distances in all directions, then the sample may be said to represent a longer time period or a larger volume, or both, than the specific point at which it was collected. In such circumstances, some sources may be represented quite well by single grab samples.

When a source is known to vary with time, grab samples collected at suitable intervals and analyzed separately can document the extent, frequency, and duration of these variations. Choose sampling intervals based on the frequency with which changes may be expected, which may vary from as little as five minutes to as long as one hour or more. When the source composition varies in space rather than time, collect samples from appropriate locations.

7.7.1.2 Composite Samples

In most cases, the term "composite sample" refers to a mixture of grab samples collected at the same sampling point at different times. Sometimes the term "time-composite" is used to distinguish this type of sample from others. Time-composite samples are most useful for observing average concentrations that will be used, for example, in calculating the loading or the efficiency of a WWTP. As an alternative to the separate analysis of many samples, followed by computation of average and total results, composite samples represent a substantial saving in laboratory effort and expense. For these purposes, a composite sample representing a 24-hour period is considered standard for most determinations. Under certain circumstances, however, a composite sample representing one shift, or a shorter time period, or a complete cycle of a periodic operation, may be preferable.

It is desirable, and often essential, to combine individual samples in volumes proportional to flow. A final sample volume of 2–31 is enough for routine analysis of WW.

7.8 Operation Records and Daily Log Sheet

All equipment operation including ON/OFF timings must be noted regularly. In addition, all voltage/amperage readings for main incomer as well as for individual motor drives must be noted hourly. The readings in all instruments such as flowmeters and pressure gauges must be noted down every two hours. Special log sheets should be prepared for this purpose.

In addition, a daily report as well as individual shift report on all aspects of the ETP must be maintained. Chemical consumption records must be updated daily. Any operating problems/special features must be noted down in detail. Tailor-made log sheets must be prepared at site to suit site conditions in which all the above details can be noted on a shift-wise as well as daily basis.

7.9 Microbiology Fundamentals

Basic to the design of a biological treatment process, or to the selection of the type of process used, is an understanding of the form, structure, and biochemical activities of the important microorganisms.

7.9.1 Basic Concepts

Microorganisms are usually grouped into three kingdoms, as shown in Table 7.12.

Table 7.12 Three kingdoms of microorganisms

Animal	Rotifers and crustaceans	Microorganisms classification
Plant	Mosses, ferns, seed plants	Multicellular with tissue differentiation
Protista Higher	Algae, protozoa, fungi, slime molds	Unicellular or multicellular without tissue differentiation
Lower	Blue-green algae, bacteria	Unicellular or multicellular without tissue differentiation

7.9.2 Cell Structure

In general, most living cells are quite similar. The interior of the cell contains a colloidal suspension of proteins, carbohydrates, and other complex compounds, called the cytoplasm. Each cell contains nucleic acids, the hereditary material that is vital to reproduction. The cytoplasmic area contains ribonucleic acid (RNA), whose major role is in the synthesis of proteins. Also within the cell wall is the area of the nucleus, which is rich in deoxyribonucleic acid (DNA). DNA contains all the information necessary for the reproduction of all the cell components and may be the blueprint of the cell.

7.9.3 Energy and Nutrient Sources

To continue to produce and function properly, an organism must have a source of energy and carbon for the synthesis of new cellular material. Inorganic elements, such as nitrogen and phosphorus, and other trace elements, such as sulfur, potassium, calcium, and magnesium, are also vital to cell synthesis. Two of the most common sources of cell carbon for microorganisms are carbon dioxide and organic matter. If an organism derives its carbon from carbon dioxide it is called autotrophic, and if it uses organic carbon it is called heterotrophic.

Energy is also needed in the synthesis of new cellular material. For autotrophic organisms the energy can be supplied by the sun, as in photosynthesis, or by an inorganic oxidation–reduction reaction. If the energy is supplied by the sun, the organism is called autotrophic photosynthetic, and if by an inorganic oxidation–reduction reaction, it is called autotrophic chemosynthetic. For heterotrophic organisms the energy needed for cell synthesis is supplied by the oxidation or fermentation of organic matter.

7.9.4 Aerobic and Anaerobic Metabolism

Organisms can also be classed according to their ability to use oxygen. Aerobic organisms can exist only when there is a supply of molecular oxygen. Anaerobic organisms can exist only in an environment devoid of oxygen. Facultative organisms can survive with or without free oxygen.

The following microorganisms are encountered in biological treatment processes: bacteria, fungi, algae, protozoa, rotifers, crustaceans, and viruses. The most important microorganisms are bacteria. Bacteria are single-cell protists that use soluble food and are found wherever moisture and a food source are available. Their usual mode of reproduction is by binary fission, although some species reproduce sexually or by budding.

Even though there are thousands of different species of bacteria, their general form falls into one of three categories: spherical, cylindrical, or helical. Representative sizes are 0.5–1.0 μm in diameter for the spherical, 0.5–1.0 μm in width by 1.5–3.0 μm in length for the cylindrical, and 0.5–5 μm in width by 6–15 μm in length for the helical.

Tests on several bacteria indicate that they are about 80% water and 20% dry material, of which 3% is organic and 10% inorganic. An approximate formula for the organic portion is $C_5H_7O_2N$. As indicated by the formula, about 53% by weight of the organic portion is carbon. The formulation $C_{60}H_{87}O_{23}N_{12}P$ has been proposed when phosphorus is also considered. Compounds comprising the inorganic portion include P_2O_5 (50%), SO_3 (15%), Na_2O (11%), CaO (9%), MgO (8%), K_2O (6%), and Fe_2O_3 (1%). Since all these elements and compounds must be derived from the environment, a shortage of any of these substances would limit and, in some cases, alter growth.

Temperature and pH play a vital role in the life and death of bacteria, as well as in other microscopic plants and animals. It has been observed that the rate of reaction for microorganisms increases with increasing temperature, doubling with approx. every 10 °C rise in temperature until some limiting temperature is reached. The pH of a solution is also a key factor in the growth of organisms. Most organisms cannot tolerate pH levels above 9.5 or below 4.0. Narrowly, the optimum pH lies between 6.5 and 7.5.

7.9.5 Role of Enzymes

The process by which microorganisms grow and obtain energy is complex and intricate; there are many pathways and cycles. Vital to the reactions involved in these pathways and cycles are the actions of enzymes, which are organic catalysts produced by the living cells. Enzymes are proteins or proteins combined with an inorganic molecule or with a low molecular weight organic molecule. As catalysts, enzymes have the capacity to increase the speed of chemical reactions greatly without altering themselves. There are two general types of enzymes: extracellular and intracellular. When the substrate or nutrient required by the cell is unable to enter the cell wall, the extracellular enzymes convert the substrate or nutrient into a form that can be transported into the cell. Intracellular enzymes are involved in the synthesis and energy reactions within the cell.

Enzymes are known for their high degree of efficiency in converting substrate to end products. One enzyme molecule can convert many molecules of substrate per minute to end products. Enzymes are also known for their high degree of substrate specificity. This high degree of specificity means that the cell must produce a different enzyme system for every substrate it uses.

The activity of enzymes is substantially affected by pH and temperature, as well as by substrate concentration. Each enzyme has a particular optimum pH and temperature. The optimum pH and temperature of the key enzymes in the cell are reflected in the overall pH and temperature preferences of the cell.

7.9.6 Dissimilatory and Assimilatory Processes

Along with enzymes, energy is required to carry out the biological reactions in the cell. Energy is released in the cell by oxidizing organic or inorganic matter or by a photosynthetic

Table 7.13 Simplified biochemical exothermic reactions for autotrophic and heterotrophic bacteria

Biochemical energy reaction	Nutrition of bacteria
$C_6H_{12}O_6 + 6O_2 \rightarrow 6CO_2 + 6H_2O$	Heterotrophic, aerobic
$C_6H_{12}O6 \rightarrow 3CH_4 + 3CO_2$	Heterotrophic, anaerobic
$2NH_4 + 3O_2 \rightarrow 2NO_2 + 2H_2O + 4H$	Autotrophic, chemosynthetic, aerobic
$5S + 2H_2O + 6NO_3 \rightarrow 5SO_4 + 3N_2 + 4H$	Autotrophic, chemosynthetic, anaerobic

reaction. The energy released is captured and stored in the cell by certain organic compounds. The most common storing compound is adenosine triphosphate (ATP). The energy captured by this compound is used for cell synthesis, maintenance, and motility. When ATP is involved in cell synthesis and maintenance, it changes to a discharged state called adenosine diphosphate (ADP). This ADP molecule can then capture the energy released in the breakdown of organic or inorganic matter. Having done this the compound again assumes an energized state as the ATP molecule. In this context, dissimilatory processes may be those processes associated with the production and/or capture of energy, whereas assimilatory processes are those associated with the production of cell tissue.

Simplified biochemical exothermic reactions for autotrophic and heterotrophic bacteria are shown in Table 7.13. A discussion of the pathways and cycles in these energy-releasing mechanisms is beyond the scope of this manual; however, in simple terms, the overall metabolism of bacterial cells can be thought of as consisting of two biochemical reactions: energy and synthesis. The first reaction releases energy so that the second reaction of cellular synthesis can proceed. Both reactions are the result of numerous systems within the cell, and each system consists of many enzyme-catalyzed reactions. The energy released in the "energy reaction" is captured by the enzyme-catalyzed system involving ATP and then transferred via ATP to the energy requiring "synthesis reaction." For heterotrophic bacteria, only a portion of the organic wastes is converted into end products. The energy obtained from this biochemical reaction is used in the synthesis of the remaining organic matter into new cells. As the organic matter in the WW becomes limiting, there will be a decrease in cellular mass, because of the utilization of cellular material without replacement. If this situation continues, all that will remain of the cell is a relatively stable organic residue. This overall process of a net decrease in cellular mass is termed endogenous respiration.

7.10 Biological Wastewater Treatment Factors

Biological WW treatment reactors are of many kinds and designs. These can be characterized by the nature of WW flow, kind of processes, and bacterial WW contact system as follows:

WW flow

- Batch.
- Continuous flow stirred tank reactor.
- Plug flow reactor.
- Arbitrary flow.

Process type

- Aerobic.
- Anerobic.
- Facultative.
- Combined anerobic and aerobic.
- Combined aerobic and anoxic.

Bacteria WW contact

- Suspended growth systems.
- Attached growth systems.
- Packed beds.
- Fluidized bed.

Important factors for selection of bioreactor design

- Process applicability.
- Applicable flow range.
- Applicable flow variation.
- Influent WW characteristics.
- Inhibiting and unaffected constituents.
- Climatic constraints.
- Reaction kinetics and reactor selection.
- Performance.
- Treatment residuals.
- Sludge handling constraints.
- Environmental constraints.
- Chemical requirements.
- Energy requirements.
- Other resource requirements.
- Reliability.
- Complexity.
- Ancillary processes required.
- Compatibility.

7.11 Lab Testing Activity and Support

The lab testing facility should include a team of chemists, microbiologists, and chemical analysts. All equipment for analysis must be made available. The quantity of equipment required can vary from lab to lab, but a minimum quantity is suggested in Table 7.14.

Table 7.14 Minimum quantity of equipment for a lab testing facility.

Description	Quantity
Comparator test set for residual chlorine of chloroscope	1 no.
Single/multiparameter instrument for pH, DO, and ammonia	1 no.
Mains-operated pH meter completed with one calomel electrode and glass electrode	1 no.
Portable turbidity meter supplied with standard accessories and user manuals	1 no.
UV-VIS spectrophotometer DR	1 no.
Water-bath	1 no.
Hot plates	1 no.
Ultrapure water plant	1 no.
Refrigerator (280 l capacity, double door)	1 no.
Muffle furnace	1 no.
Magnetic stirrer	1 no.
Analytical balance (electronic) with weight box, capacity 220 g, resolution 0.0001	1 no.
Jar test apparatus-6 stirrer	1 no.
Centrifuge	1 no.
Field kit (to measure pH, pocket-type)	1 no.
TOC analyser	1 no.
Hot air oven	1 no.
Autoclave	1 no.
Cotton/aluminum foils	1 pack
Burners (bunsen) with pilot lamp	1 no.
Suction flask (1 l capacity)	1 no.
Suction pump	2 nos
Sampling/wash bottles	2 nos
Measuring cylinders (1000, 500, 250, 100, 50, 25, 10, and 5 ml)	2 nos each
Volumetric flasks (500, 250, 100, and 50 ml)	2 nos each
Pipette (20, 10, 5, and 2 ml)	2 nos each
Burette (50 ml with stand)	2 nos
Funnel	2 nos
Whatman filter paper	5 pack
Normal grade 1 filter paper	2 pack
Vacuum pump	1 no.
Soxhlet extraction unit	1 no.
Kjeldahl digestion unit	1 no.
Weighing balance (max. 10 kg)	1 no.

(Continued)

Table 7.14 (Continued)

Description	Quantity
Beakers (500, 100, and 50 ml)	2 nos each
Desiccator	2 nos
Gooch/silica crucible	2 nos
Moisture dishes	2 nos
Round-bottom flask (500, 250, 100, and 50 ml)	2 nos each
Pipette stand	1 no.
Kjeldahl flask	2 nos
Filtration assembly for suspended solid	1 no.
Incubator (37 or 44 °C)	1 no.
Tongue	2 nos
Hand gloves	5 sets
Miscellaneous as per specific lab requirement	1 no.

7.12 Best Practices for Pipe Line Sizing

For primary treatment, 1-inch pipes are considered the minimum size for dosing lines and sludge handling lines. This is based on calculation in view of chocking and flushing issues. The flow velocities in Table 7.15 can be considered for better line size calculation.

Table 7.15 Flow velocity for pump suction and pump discharge.

Type of service	Flow velocity (m/s)	
	Pump suction	Pump discharge
Wastewater	0.6–0.1	1.2–1.8
Slurry and sludge	0.5–0.8	1.0–1.2

7.13 Best Practices for Instrument Operation

For good operation of instrument equipment, some guidelines for analysers, level control, and flow and pressure measurement equipment are given below:

- Service/process fluid.
- Contact and non-contact basis.
- Local field indication.
- Transmitter type.

7.14 Wastewater Online Monitoring Process

The WW online monitoring process depends on type of industry and treatment process being adopted for raw effluent treatment. With rapid industrialization, it is becoming a need and necessity to regulate and minimize inspection of industries on a routine basis. Therefore efforts need to be made to bring self-discipline in the industries to exercise self-monitoring and compliance and transmit data of effluent and emission to regulatory bodies on a continuous basis.

For strengthening monitoring and compliance through the self-regulatory mechanism, online emission and effluent monitoring systems need to be installed and operated by the industries.

7.14.1 Advantages of Online Monitoring Techniques

The major advantages of online monitoring systems over traditional laboratory based and portable field methods are:

- Online monitoring systems provide continuous measurement of data for long periods of time, at the monitoring site of interest, without skilled staff being required to perform the analysis.
- All the major steps in traditional analysis like sample collection, transportation, conditioning, calibration, and analysis procedures including quality control (QC) are usually automated in the sampling systems and online analysers.
- In case of sudden disturbance in the production process/pollution control system, the online analysers provide timely information for taking immediate corrective/preventive steps compared to conventional methods.

Installation of an online effluent quality monitoring system at the outlet of a WWTP for measurement of parameters pH, COD, and BOD and other industry-specific parameters is suggested (see Table 7.16).

Table 7.16 Online monitoring industry-specific parameters.

Industry category	Effluent parameters
Aluminum	pH, BOD, COD, TSS, flow
Distillery	pH, BOD, COD, TSS, flow
Dye and dye intermediate	pH, BOD, COD, TSS, chromium, flow
Chlor alkali	pH, TSS, flow
Fertilizers	pH, flow, ammoniacal nitrogen, fluoride
Iron and steel	pH, phenol, cyanide, flow
Oil refinery	pH, BOD, COD, TSS, flow
Petrochemical	pH, BOD, COD, TSS, flow
Pesticides	pH, BOD, COD, TSS, chromium, arsenic, flow
Pharmaceuticals	pH, BOD, COD, TSS, chromium, arsenic, flow

(Continued)

Table 7.16 (Continued)

Industry category	Effluent parameters
Power plants	pH, TSS, temperature
Pulp and paper	pH, BOD, COD, TSS, AOx, flow
Sugar	pH, BOD, COD, TSS, flow
Tannery	pH, BOD, COD, TSS, chromium, flow
Zinc	pH, TSS, flow
Copper	pH, TSS, flow
Textile	pH, COD, TSS, flow
Dairy	pH, BOD, COD, TSS, flow
Slaughter house	pH, BOD, COD, TSS, flow

7.15 Wastewater Characteristics Monitoring Parameters

The WW lab monitoring process depends on type of industry and treatment process being adopted for raw effluent treatment. However, WW characteristics parameters that need to be tested on a daily basis for inlet and outlet of effluent are listed in Table 7.17.

- Ensure daily sampling and analysis of influent to ETP plant for parameters pH, COD, TSS, MLSS, MLVSS, ammonical nitrogen, etc.
- Measure the outlet flow hourly and note in the logbook.

Table 7.17 Wastewater characteristics parameters.

Wastewater parameters	Units
pH	
Temperature	°C
Color	Pt-Co
TSS	mg/l
TDS	mg/l
BOD (3 days)	mg/l
COD	mg/l
Oil and grease	mg/l
Phenolic compounds	mg/l
Sulfides	mg/l
Ammoniacal nitrogen	mg/l
Chloride	mg/l

Table 7.17 (Continued)

Wastewater parameters	Units
Sulphate	mg/l
Cyanide	mg/l
Fluoride	mg/l
Hexavalent chromium	mg/l
Total chromium	mg/l
Copper	mg/l
Nickel	mg/l
Zinc	mg/l
Mercury	mg/l
Lead	mg/l
Arsenic	mg/l
Cadmium	mg/l
Bioassay test	90% survival of fish after 96 h in 100% effluent

7.16 Effluent Treatment Plant Operating Procedure

Table 7.18 shows the operating procedure for ETPs.

Table 7.18 Effluent treatment plant operating procedure.

Unit operation	Details of operation	Remarks
Equalization tank	To equalize/control influent flow of WW	Ensure proper operation of aeration at equalization to ensure proper mixing of influent
Neutralization tank	Neutralize influent received from equalization tank	Ensure optimum dosage of lime required for neutralization
Flash mixer	Mixing of alum with WW for better result of process	Ensure proper dose of alum and operation of flash mixer (rpm) to achieve maximum floc generation for removal of suspended solids
Flocculator	Presedimentation stage of treatment by settlement of coarse particles	Addition of optimum dose of polyelectrolyte for bonding of flocs
Primary clarifier	Fine particles settle to bottom of tank by zone settling process	Retention time should be maintained for three to four hours
Bioreactor	BOD/COD reduction is achieved by conventional biological treatment	Ensure proper air supply, pH 6.5–8, and DO 3–4 mg/l in aeration tank through diffusers
		Maintain MLSS between 4000 and 6000 mg/l by addition of external food or wasting of sludge

(Continued)

Table 7.18 (Continued)

Unit operation	Details of operation	Remarks
Secondary clarifier	Settlement of biomass and degraded organic compounds	Ensure proper recirculation of biomass to aeration tank. Sludge generated should be recycled or wasted as per MLSS/MLVSS levels
Guard pond	Storage of treated WW with high retention time	Only storage of treated WW. Ensure good housekeeping in area
Treated WW tank	Treated WW should be discharged as per regulatory norms	Ensure continuous online monitoring of treated WW quality and quantity
Sludge treatment	Dewatering of sludge	Regular cleaning and disposal of wet sludge cakes and water should be sent to primary clarifier

Further Reading

American Public Health Association (1985). *Standard Methods for the Examination of Wastewater*, 16e. Washington, DC: APHA, AWWA, WPCE.

Byrne, W. (2002). *Reverse Osmosis: A Practical Guide for Industrial Users*, 2e, 179. Tall Oaka Publishing Inc.

Chaubey, M. (2006a). Fouling prevention techniques of reverse osmosis systems. *Water Digest* I (1).

Chaubey, M. (2006b). Chemical cleaning techniques of reverse osmosis systems. *Water Digest* 2 (2).

Chaubey, M. (2019). Best practices & design considerations for wastewater treatment with MBBR technology. *Official Journal of Indian Chemical Council*: 16–19.

CPHEEO (2013). *Manual on Sewerage and Sewage Treatment Systems – 2013*. New Delhi: Central Public Health & Environmental Engineering Organization, Ministry of Urban Development.

Hrenovic, J., Buyukgungor, H., and Orhan, Y. (2003). Use of natural zeolite to upgrade activated sludge process. *Food Technology* 41: 157–165.

Metcalf & Eddy Inc (2014). *Wastewater Engineering: Treatment and Resource Recovery*, 5e. New Delhi: Tata McGraw-Hill.

Glossary

Activated carbon A highly adsorptive material used to remove organic substances from water.

Activated silica A coagulant aid used to form a denser, stronger floc.

Aerobic A process that takes place in the presence of air or oxygen.

Algae Primitive plants, one- or many-celled, usually aquatic and capable of photosynthesis.

Alum The most common chemical used for coagulation. It is also called aluminum sulfate.

Anerobic A process that takes place without air or oxygen.

Anoxic The process by which nitrate is converted biologically into nitrogen gas in the absence of oxygen.

Aquatic Living in water.

Aquifer A formation or group of formations or part of a formation that contains sufficient saturated permeable material to yield economical quantities of water to wells and springs.

Artesian well A well deriving its water from a confined aquifer in which the water level stands above the ground surface. It can also be water that is forced from the aquifer by compaction caused by the weight of overlying sediments.

Biodegradable Capable of being broken down by biological processes.

Biochemical oxygen demand (BOD) The amount of oxygen required to oxidize the various organic chemicals in wastewater treatment.

Bit Cutting tool attached to the bottom of the drill stem.

Capillary fringe The zone where groundwater is drawn upward by capillary force.

Carbonate Sediment formed by the organic or inorganic precipitation from aqueous solution of carbonates of calcium, magnesium, or iron.

Carbonate rock A rock consisting of carbonate minerals, such as limestone and dolomite.

Cations An ion having a positive charge and, in electrolytes, characteristically moving toward a negative electrode.

Chemical oxygen demand (COD) A measure of the organic material in wastewater that can be oxidized chemically using dichromate in an acid solution. Normally COD values are always higher than the BOD value of any wastewater

Wastewater Treatment Technologies: Design Considerations, First Edition. Mritunjay Chaubey.
© 2021 John Wiley & Sons Ltd. Published 2021 by John Wiley & Sons Ltd.

Chlorination The process of adding chlorine to water to kill disease-causing organisms or to act as an oxidizing agent.

Chlorinator Any device that is used to add chlorine to water.

Chlorine residual The amount of chlorine present in the distribution system.

Coagulant A chemical used in water treatment for coagulation. The most common coagulants are aluminum sulfate (alum) and ferric sulfate.

Coagulant aid A chemical added during coagulation to improve the process by stimulating floc formation or by strengthening the floc so it holds together.

Coagulation The water treatment process that causes very small suspended particles to attract one another and form large particles.

Coliforms A group of bacteria, some of them fecal coliforms, normally found in human and animal feces. They grow in the presence of bile salts and ferment lactose-producing acids and gas.

Colloidal particles Extremely small solid particles that will not settle out of a solution (sizes from 0.0001 to $1\,\mu m$).

Contamination The degradation of water quality from its natural condition as a result of human and animal activities.

Detention time The average length of time a drop of water or a suspended particle remains in a tank or chamber. Mathematically, it is the volume of water in the tank divided by the flow rate through the tank.

Digestion The breaking down of organic waste by bacteria.

Disinfection The water treatment process that kills disease-causing organisms in water. Chlorine is the most common chemical used.

Dissociation The processes in which water has the natural tendency to break down part of any volume of water spontaneously into hydrogen (H+) and hydroxyl (OH−).

Dissolved solid Any material that is dissolved in water and can be recovered by evaporating the water after filtering the suspended material.

Drawdown The distance below the water table that the water table in a well falls to when steady-state pumping is in progress. It is the distance between the static water level and the dynamic water level.

Dynamic water level A water level in a well during steady-state pumping.

Effluent A waste liquid discharge from an industry or municipal treatment process in its natural state or partially or completely treated and discharged into the environment (such as into streams, rivers, lakes, and seas).

Erosion The general process or group of processes whereby the materials of the earth's crust are moved from one place to another by running water, waves and currents, wind, or glacier ice.

Filtration The water treatment process involving the removal of suspended matter by passing the water through a porous medium such as sand.

Floc Collections of smaller particles that have come together into larger particles as a result of coagulation/flocculation processes in water treatment.

Flocculation The water-treatment process following coagulation that uses gentle stirring to bring suspended particles together so they will form larger particles, or clumps called floc.

Hardness A property of water that causes an insoluble residue to form when the water is used with soap. It is primarily caused by the presence of calcium and magnesium ions.

Humic/humus Material resulting from the decay of leaves and other plant matter.

Infiltration The process in which water is seeping to the ground or entering a sewer system, including sewer service connections, from the ground, or through such means as, but not limited to, defective pipes, pipe joints, connections, or manhole walls.

Ion An element or compound that has gained or lost an electron, so that it is no longer electrically neutral but carries a charge.

Loading rate The flow-rate per unit area of a sewage, filter, or ion exchange unit.

Pathogen A disease-causing organism.

Percolate The act of water seeping or filtering through the soil without a definite channel.

Pretreatment/preliminary treatment Any physical, chemical, or mechanical process used before the main water treatment processes, such as screening, pre-sedimentation, and chemical addition.

Reverse osmosis (RO) RO is the flow of water through a semipermeable membrane from the more concentrated solution into the diluted solution under influence of pressure greater than the osmotic pressure.

Runoff Precipitated water flowing to streams and rivers.

Saturation A point at which a solution can no longer dissolve any more of a particular chemical. Precipitation of the chemical will occur beyond saturation point.

Screening A pretreatment method that uses coarse screens to remove large debris from the water to prevent clogging of pipes or channels to the treatment plant.

Sewage A waste that includes excreta and other domestic and municipal wastes and industrial effluents.

Static water level The level of water in a well that is not being affected by withdrawal of groundwater.

Transpiration The process by which water is absorbed by plants through its roots and evaporated into the atmosphere from the plant surface.

Vadose zone The zone containing water under pressure less than that of the atmosphere, including soil, water, intermediate vadose water, and capillary water. This zone is limited above the land surface and below by the surface of the zone of saturation (i.e. the water table).

Wastewater Domestic sewage, industrial effluent, or a combination of these two, as in the case of municipal sewage from industrial areas.

Waterborne disease A disease caused by a waterborne organism or toxic substance.

Water table The surface between the vadose zone and the groundwater; that surface of a body of unconfined groundwater at which the pressure is equal to that of the atmosphere.

Weathering The in-situ physical disintegration and chemical decomposition of rock materials at or near the earth's surface.

Index

Wastewater Treatment Technologies: Design Considerations, First Edition. Mritunjay Chaubey.
© 2021 John Wiley & Sons Ltd. Published 2021 by John Wiley & Sons Ltd.